Exploring AutoCAD Raster Design 2016

CADCIM Technologies

525 St. Andrews Drive
Schererville, IN 46375, USA
(www.cadcim.com)

Contributing Author

Sham Tickoo

Professor
Purdue University Calumet
Hammond, Indiana, USA

CADCIM Technologies

Exploring AutoCAD Raster Design 2016
Sham Tickoo

CADCIM Technologies
525 St Andrews Drive
Schererville, Indiana 46375, USA
www.cadcim.com

ISBN   978-1-936646-96-8

www.cadcim.com

DEDICATION

To teachers, who make it possible to disseminate knowledge
to enlighten the young and curious minds
of our future generations

To students, who are dedicated to learning new technologies
and making the world a better place to live in

SPECIAL RECOGNITION

A special thanks to Mr. Denis Cadu and the ADN team of Autodesk Inc.
for their valuable support and professional guidance to
procure the software for writing this textbook

THANKS

To employees of CADCIM Technologies for their valuable help

Online Training Program Offered by CADCIM Technologies

CADCIM Technologies provides effective and affordable virtual online training on various software packages including Computer Aided Design, Manufacturing, and Engineering (CAD/CAM/CAE), computer programming languages, animation, architecture, and GIS. The training is delivered 'live' via Internet at any time, any place, and at any pace to individuals as well as the students of colleges, universities, and CAD/CAM/CAE training centers. The main features of this program are:

Training for Students and Companies in a Classroom Setting

Highly experienced instructors and qualified engineers at CADCIM Technologies conduct the classes under the guidance of Prof. Sham Tickoo of Purdue University Calumet, USA. This team has authored several textbooks that are rated "one of the best" in their categories and are used in various colleges, universities, and training centers in North America, Europe, and in other parts of the world.

Training for Individuals

CADCIM Technologies with its cost effective and time saving initiative strives to deliver the training in the comfort of your home or work place, thereby relieving you from the hassles of traveling to training centers.

Training Offered on Software Packages

CADCIM Technologies provides basic and advanced training on the following software packages:

CAD/CAM/CAE: *CATIA, Pro/ENGINEER Wildfire (Creo), SOLIDWORKS, Autodesk Inventor, Solid Edge, NX, AutoCAD, AutoCAD LT, Customizing AutoCAD, AutoCAD Electrical, EdgeCAM, and ANSYS*

Architecture and GIS*: AutoCAD Raster Design, AutoCAD Map 3D, AutoCAD Civil 3D, Autodesk Revit Structure, Autodesk Revit Architecture, Autodesk Navisworks, Autodesk Revit MEP, STAAD.Pro, and Oracle Primavera P6*

Animation and Styling*: Autodesk 3ds Max, 3ds Max Design, Maya, Alias, Pixologic ZBrush, Character Animation, and The Foundry NukeX*

Computer Programming*: C++, VB.NET, Oracle, AJAX, and Java*

*For more information, please visit the following link: **http://www.cadcim.com.***

Note

If you are a faculty member, you can register by clicking on the following link to access the teaching resources: ***www.cadcim.com/Registration.aspx***. The student resources are available at ***www.cadcim.com***. We also provide **Live Virtual Online Training** on various software packages. For more information, write us at ***sales@cadcim.com***.

Table of Contents

Chapter 3: Image Management Tools

Chapter 4: Image Processing

Chapter 5: Raster Entity Manipulation (REM) Tools

Chapter 6: Vectorization Tools

Chapter 7: Multispectral Images and Digital Elevation Models

This page is intentionally left blank.

Preface

AutoCAD Raster Design 2016

AutoCAD Raster Design 2016, developed by Autodesk Inc., is a complete suite used for managing and editing raster data. Built on the latest release of AutoCAD software, it has various tools that help in creating, editing, analyzing, and interpreting various kinds of raster datasets effectively. AutoCAD Raster Design has a wide range of applications in GIS and Mapping, Architecture, Engineering and Construction (AEC), Mechanical industries and other industries.

AutoCAD Raster Design has interoperability with major design and data conversion software packages. This feature allows the Raster Design users to access CAD and GIS data from various sources to perform raster to vector conversion. In AutoCAD Raster Design, you can connect a raster dataset at the software platform and georeference it with ease. In addition, you can use various raster and vector manipulation, and raster to vector data analysis tools to create customized datasets based on your project requirements. AutoCAD Raster Design also allows you to import different raster images of multiple color tones and scanned CAD drawings or maps. Moreover, you can import Multispectral Remote Sensing images, Digital Elevation Models, and Digital Terrain Models in AutoCAD Raster Design to analyse and perform slope analysis and also create slope maps. AutoCAD Raster Design provides you with the powerful tools for displaying and publishing vectorized data.

Exploring AutoCAD Raster Design 2016 is a comprehensive textbook that has been written to cater to the needs of the students and the professionals. The chapters in this textbook are structured in a pedagogical sequence, which makes the learning process very simple and effective for both the novice as well as the advanced users of AutoCAD Raster Design. In this textbook, complex vectorization processes have been illustrated through easy-to-understand flow diagrams. Also, various processes such as manipulating and managing old CAD data and displaying spatial data have been covered in this textbook. This edition also introduces users to the concepts of industry model database for managing spatial data. The simple and lucid language used in this textbook makes it a ready reference for both the beginners and the intermediate users.

The salient features of the textbook are as follows:

*   **Tutorial Approach**
    The author has adopted the tutorial point-of-view and learn-by-doing approach throughout the textbook. This approach guides the users through various processes involved in creating and analyzing spatial data. At the end of each chapter, tutorials are provided to practice the concepts learned in the chapter.

- **Real-World Projects as Tutorials**
  The author has used about 13 real-world GIS projects as tutorials in this book. This will enable the readers to relate the tutorials to the real-world projects in GIS industry. In addition, there are about 10 exercises based on the real-world GIS projects.

- **Tips and Notes**
  The additional information related to various topics is provided to the users in the form of tips and notes.

- **Learning Objectives**
  The first page of every chapter summarizes the topics that are covered in that chapter.

- **Self-Evaluation Test, Review Questions, and Exercises**
  The chapters end with Self-Evaluation Test so that the users can assess their knowledge of the chapter. The answers to Self-Evaluation Test are given at the end of the chapter. Also, the Review Questions and Exercises are given at the end of the chapters and they can be used by instructors as test questions and exercises.

- **Heavily Illustrated Text**
  The text in this book is heavily illustrated with about 200 line diagrams and screen capture images.

Symbols Used in the Textbook

Note
The author has provided additional information about some topics in the form of notes.

Tip
The author has provided extra information to the users about the topic being discussed in the form of tips.

Formatting Conventions Used in the Textbook

Please refer to the following list for the formatting conventions used in this textbook.

- Names of tools, buttons, options, browser, palette, panels, and tabs are written in boldface

  Example: The **Rubber Sheet** tool, the **Rub** radio button, the **Edit** panel, the **Raster Tools** tab, **Properties** palette, and so on.

- Names of dialog boxes, drop-downs, drop-down lists, list boxes, areas, edit boxes, check boxes, and radio buttons are written in boldface.

  Example: The **AutoCAD Raster Design Options** dialog box, the **Units** drop-down list, the **Density** edit box in the **Image Defaults** tab, and so on.

- Values entered in edit boxes are written in boldface.

  Example: Enter **2** in the **Float tolerance (pixels)** edit box.

- Names of the files saved are italicized.

  Example: *c03_Tut01_Result.dwg*

- The methods of invoking a tool/option from the ribbon, Application Menu, or the shortcut keys are given in a shaded box.

Ribbon:	Raster Tools > Insert & Write > Insert
Command:	IINSERT

Naming Conventions Used in the Textbook

Tool

If you click on an item in a panel of the ribbon and a command is invoked to create/edit an object or perform some action, then that item is termed as **tool**.
For example: **Polyline Follower** tool and **Bias** tool

If you click on an item in a panel of the ribbon and a dialog box is invoked wherein you can set the properties to create/edit an object, then that item is also termed as **tool**, refer to Figure 1.
For example: **Insert** tool and **Rubber Sheet** tool

Figure 1 Tools in the Ribbon

Button

The item in a dialog box that has a 3d shape like a button is termed as **Button**. For example, **OK** button, **Cancel** button, **Apply** button, and so on, refer to Figure 2. If the item in a ribbon is used to exit a tool or a mode, it is also termed as button. For example, **Options - Image Default** button, **Open** button, and so on.

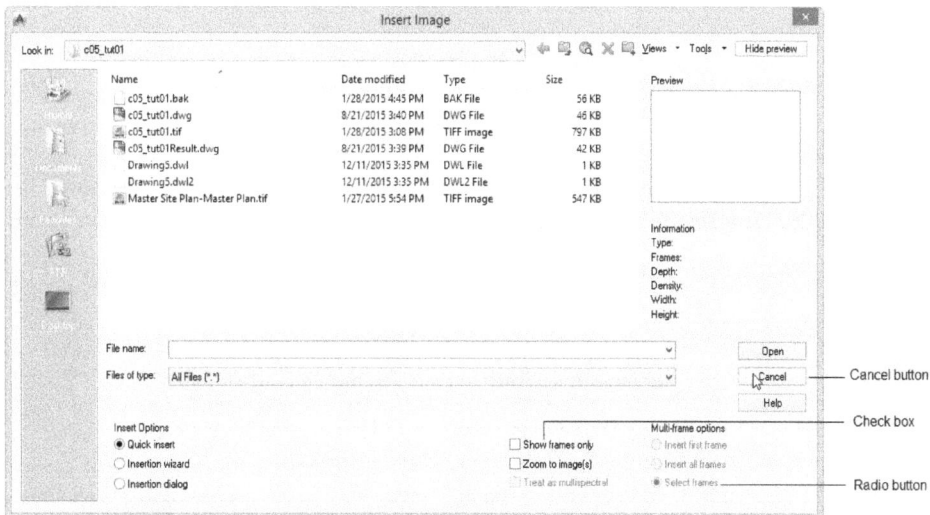

*Figure 2 Choosing the **Cancel** button*

Drop-down

A drop-down is the one in which a set of common tools are grouped together. You can identify a drop-down with a down arrow on it, refer to Figure 3. These drop-downs are given a name based on the tools grouped in them. For example, **Followers** drop-down, **Crop** drop-down, and so on.

Figure 3 *Choosing a tool from a drop-down*

Drop-down List

A drop-down list is the one in which a set of options are grouped together. You can set various parameters using these options. You can identify a drop-down list with a down arrow on it, refer to Figure 4. For example, **Contour creates** drop-down list, **Vertical Units** drop-down list, and so on.

Figure 4 *Selecting an option from the **Contour creates** drop-down list*

Options

Options are the items that are available in shortcut menus, dialog boxes, drop-down lists, and so on. For example, choose the **Browse path** option from the shortcut menu displayed, refer to Figure 5.

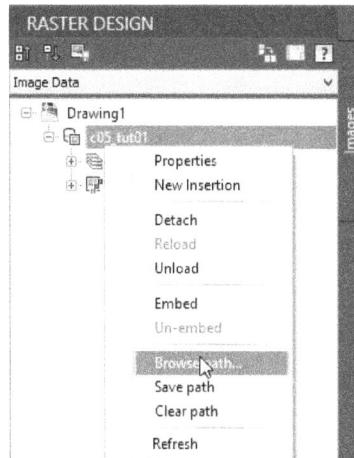

Figure 5 *Choosing an option from the shortcut menu*

Free Companion Website

It has been our constant endeavor to provide you the best textbooks and services at affordable price. In this endeavor, we have come out with a Free Companion website that will facilitate the process of teaching and learning of AutoCAD Raster Design 2016. If you purchase this textbook, you will get access to the files on the Companion website.

Faculty Resources

- **Technical Support**
  You can get online technical support by contacting ***techsupport@cadcim.com***.

- **Instructor Guide**
  Solutions to all review questions and exercises in the textbook are provided in the instructor guide to help the faculty members test the skills of the students.

- **PowerPoint Presentations**
  The contents of the book are arranged in PowerPoint slides that can be used by the faculty for their lectures.

- **Part Files**
  The Raster Design files used in illustration, tutorials, and exercises are available for free download.

Student Resources

- **Technical Support**
  You can get online technical support by contacting ***techsupport@cadcim.com***.

- **Part Files**
  The Raster Design files used in illustrations and tutorials are available for free download.

If you face any problem in accessing these files, please contact the publisher at *sales@cadcim.com* or the author at *stickoo@purduecal.edu* or *tickoo525@gmail.com*.

Stay Connected

You can now stay connected with us through Facebook and Twitter to get the latest information about our textbooks, videos, and teaching/learning resources. To stay informed of such updates, follow us on Facebook (*www.facebook.com/cadcim*) and Twitter (*@cadcimtech*). You can also subscribe to our YouTube channel (*www.youtube.com/cadcimtech*) to get the information about our latest video tutorials.

Chapter *1*

Introduction to AutoCAD Raster Design 2016

Learning Objectives

After completing this chapter, you will be able to:

- *Understand raster data*
- *Understand features of AutoCAD Raster Design*
- *Understand applications of AutoCAD Raster Design*
- *Configure options in AutoCAD Raster Design*

INTRODUCTION

AutoCAD Raster Design comes up as a solution for working with raster data within the Autodesk product family. AutoCAD Raster Design is an add-on product and can be used along with Autodesk products such as AutoCAD Civil 3D and AutoCAD Map 3D. It has powerful tools for sizing, cleaning up and processing raster images into drawings.

AutoCAD Raster Design helps you to insert, manipulate, and edit raster images into a drawing. This package has various image correction tools to georeference the raster image. You can also apply various image enhancements that will aid in better data interpretation. The software contains Raster Entity Manipulation (REM) tools to vectorize raster entities and then manipulate them. The software also contains Optical Character Recognition (OCR) tools that can convert an image text directly into an editable AutoCAD text entity. Also, you can modify and edit any CAD drawing saved in *.pdf format. This package allows you to convert raster geometries into vector geometries which can be saved and used for further references.

UNDERSTANDING RASTER DATA

Raster data consists of spatially coherent and digitally stored numerical data collected from satellite sensors, aerial cameras, and scanners. A digital image can be represented as arrays of pixels and each pixel corresponding to a digital number (DN) representing the brightness level of that pixel. These type of digital images are referred to as raster data in which pixels (cells) are organized into rows and columns. Figure 1-1 shows the raster data acquired from a remote sensing satellite.

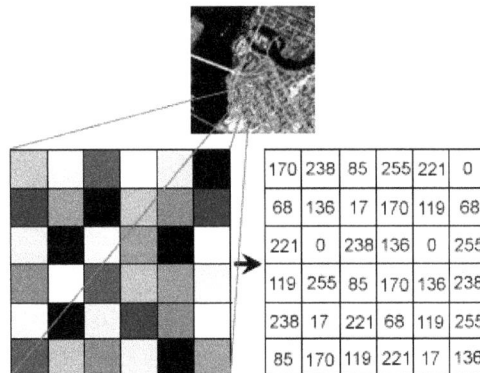

Figure 1-1 Raster data acquired from a remote sensing satellite

Raster data, such as aerial photographs and satellite images is a source of data commonly used to collect information in various remote sensing applications. Rasters are also used to collect, store, and display information of the spatial entities in a digital format. A raster data may be a source of continuous or discrete information. For example, a thematic map is used to represent discrete data such as vegetation type or complete land use/land cover. Similarly, rasters are also used to represent continuous data such as elevation and temperature. Raster data is a source of information which complement vector data in many GIS applications. Figure 1-2 shows a thematic map of the land use/land cover classification for a geographical region representing the correlation of satellite images with GIS applications.

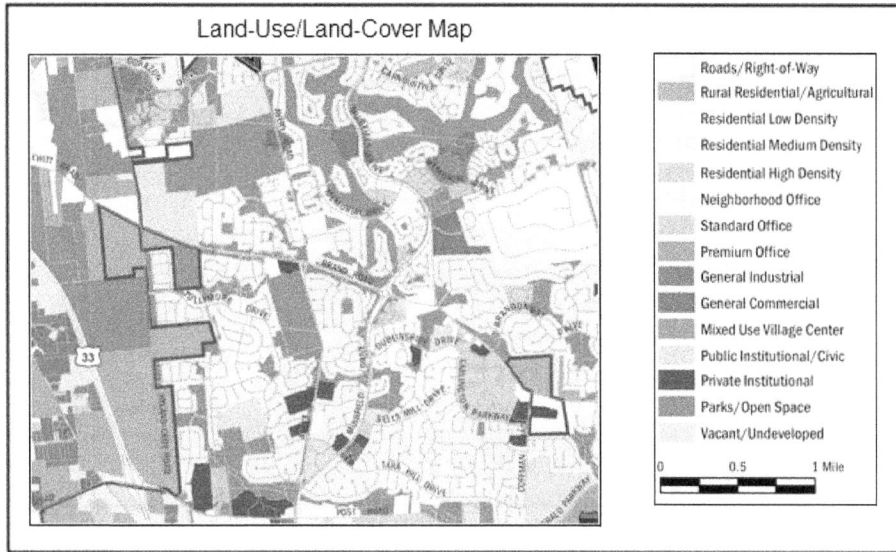

Figure 1-2 The Land-Use/Land-Cover Map

Properties of Raster Data

A raster image has various properties such as number of pixels, band (layer), cell size, data type, and so on. Some rasters, such as the digital elevation models and satellite images also have geographic properties that define the correlation parameters of an image. These properties of raster image govern its behavior and information. Some of the raster properties are briefly discussed next.

Cell Size

Cell size is the level of detail of features represented by a raster. Raster data divides a given scene into grid cells or pixels. Each grid cell is composed of cell attributes which is represented by certain values also referred to as pixel values or grid values or Digital Number (DN) values, and are arranged in a grid (matrix). The size of the raster is determined by its cell size. Smaller cell size results in larger raster dataset and higher resolution whereas larger cell size results in lower resolution, refer to Figure 1-3.

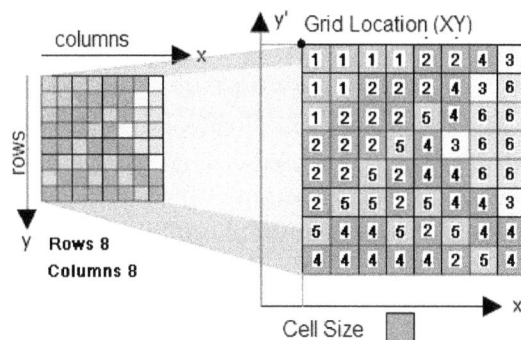

Figure 1-3 Height and width of a raster

Data Type

Every cell in the raster contains a value that represents information. For example, in an elevation map, the value of a cell represents the height of the terrain. The value stored in the raster cell is mostly numeric and can have a data type such as 1 bit integer, 2 bit integer, 4 bit integer, 8 bit integer, 16 bit integer, 32 bit integer, and more. Depending upon the intensity of the values, raster can be classified into three types: bitonal (mono color), grayscale, and multispectral.

Note
Due to the simple data structure, arithmetic operations can be easily performed on a raster data. As a result, raster datasets are often used in Geographical Information System (GIS) as a data source for performing spatial analysis.

Bitonal Rasters
Bitonal rasters are also called binary images or mono color. In these type of images, cells are represented in black or white color. You can store the value of the cell using a single bit. If the value of cell is 0, it is rendered black and if the value of the cell is 1, it is rendered white.

Grayscale Rasters
A gray scale color model is used to display raster in different shades of gray. A grayscale image is the one in which Red, Green and Blue have equal intensity in RGB space .

Multispectral Rasters
A color raster contains pixels rendered in multiple colors such as a colored photograph. Figure 1-4 shows a bitonal, a grayscale, and a colored image.

| Bitonal Raster | Grayscale Raster | Multispectral Raster |

Figure 1-4 Bitonal, grayscale, and multispectral image

Bands

A raster dataset contains one or more layers called bands. A band is represented by a single matrix of cell values. Some rasters are composed of single band, for example, digital elevation model and some contain multiple bands such as multispectral, hyperspectral, and ultraspectral images. In case of a multiband raster, the matrix of cells in every band spatially coincide with the matrix in other band.

Geographical Properties

The geographic properties of a raster define its correlation parameters. These correlation parameters are important when the rasters are used as data sources in GIS. Using these correlation parameters you can georeference a raster. The geographic properties of a raster include information such as the X and Y coordinates of the origin, cell size, and number of rows and columns in the raster. The cell size also represents the spatial resolution of the raster. This parameter defines the actual area on the ground that is represented by one cell (pixel) in the raster.

FEATURES OF AutoCAD RASTER DESIGN

AutoCAD Raster Design is an add-on product developed by Autodesk for the professionals who intend to use raster data in their project. This application enables the users to manage, convert, analyze, and edit raster data in various types and formats.

Note

AutoCAD Raster Design is an add-on product and therefore requires AutoCAD based application for its installation such as AutoCAD Map3D or AutoCAD Civil 3D.

Some of the basic features of AutoCAD Raster Design are discussed next.

Raster Management

AutoCAD Raster Design provides some enhanced tools along with the basic tools for image management. These tools enable the users to work with raster data of various file formats, such as Digital Elevation Model (DEM) and multispectral satellite images. AutoCAD Raster Design also provides tools for georeferencing and transforming coordinate system of the raster images.

Image Enhancement and Display

AutoCAD Raster Design can improve the appearance of an image by adjusting tone, brightness and contrast, equalize histogram, and by applying filters. The image processing tools provide various options for performing such adjustments. The image enhancement feature is useful for performing image interpretation such as feature identification and data convention.

This software also contains palette manager that provides the flexibility of choosing the color palette for displaying images. Using this palette manager, a user can combine, change, and remove colors from an image.

Image Cleanup and Editing

AutoCAD Raster Design is also equipped with a set of image cleanup tools. Using these image cleanup tools in AutoCAD Raster Design application, you can rectify and remove errors such as skew and speckles from an image. Furthermore, you can edit the raster entities in an image. The **Touchup** tool is used to clean scanned drawings and maps. The Raster Entity Manipulation (REM) tools are effective raster editing tools. Using the REM tools, you can perform editing tasks such as creating, deleting, trimming, and extending raster entities. The usage of the REM tools for editing raster is similar to the usage of editing tools in AutoCAD.

This software also provides the raster snapping options. These options provide various snapping modes that can be used to carry out precise raster editing operations.

Raster to Vector Conversion Tools

In AutoCAD Raster design, you can convert raster data into vector data using the vectorization tools. AutoCAD Raster Design uses the smart correct settings to increase the precision with which the vectors are created. These tools provide advance features such as dynamic dimensioning and grip editing. While converting data from raster to vector, the Optical Character Recognition (OCR) feature of AutoCAD Raster Design identifies the text in the raster and converts it into a vector data. This eliminates the need of retyping the text thereby saving considerable amount of time and effort. The OCR feature of AutoCAD Raster Design can only identify the text that is hand written or machine typed and is optimized for images with a resolution of 300 dpi (dots per inch).

Note
The OCR is optimized for images with a resolution of 300 dpi. Undesirable results may be produced if used with images having resolution below 150 dpi or above 400 dpi.

Data Analysis

AutoCAD Raster Design can work with raster data such as multispectral satellite imagery. Using this software, you can combine data from different bands of a multispectral image to compose a false color image. This data can be further used to analyze information such as land use and land cover. You can also use the AutoCAD Raster Design to perform elevation, slope, and aspect analysis using the Digital Elevation Models. Raster query tools also help to find the unique values from raster datasets.

These features make AutoCAD Raster Design a versatile tool with applications in various fields.

APPLICATIONS OF AutoCAD RASTER DESIGN

Professionals such as GIS analysts, Civil engineers, Utility engineers, Architects, Town planners, and CAD draftsmen extensively work with raster data in their day-to-day activities. AutoCAD Raster Design has been developed considering the needs of these professionals to frequently use the raster data. The applications of AutoCAD Raster Design in various industries are discussed next.

GIS and Mapping Industry

GIS project often use aerial photographs, satellite images, and scanned copies of old maps as data sets for creating base maps. These data sets can be acquired from different sources and may have different coordinate systems. These data also come in a variety of file formats. AutoCAD Raster Design supports a wide range of file formats including *.grid*, *.ascii*, *.dem*, and *.tiff* of satellites such as QuickBird and Landsat series. It is also capable of transforming the coordinate system of the images to match the current coordinate system of the drawing. In AutoCAD Raster Design, image mosaicing helps the mapping professionals in managing large amount of data. This software can be used to combine various bands of a multispectral satellite imagery to compose a false color image. Figure 1-5 shows a Landsat ETM+ image in various band combinations. These false color images can be used to analyze data such as vegetation cover, land use and many more.

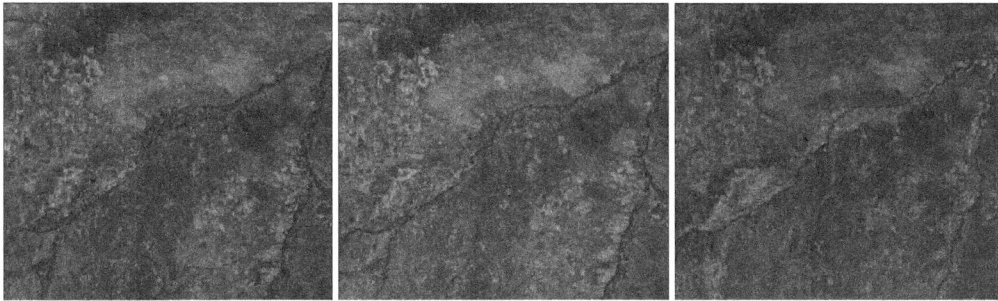

Figure 1-5 A multispectral satellite image in various band combinations

Architecture, Engineering, and Construction (AEC) Industry

The Architecture, Engineering, and Construction (AEC) Industry is another stream of professionals that work with huge volumes of raster data. This industry is primarily involved in design, construction and maintenance (operation) of various infrastructure projects. The design and construction of engineering projects such as roads, dams, and canals is governed by topographical characteristics of the terrain. As a result, the study of the topography is the first step toward designing these types of infrastructure projects.

AutoCAD Raster Design not only supports the raster data in Digital Elevation Model (DEM) format but also using it you can analyze the DEM data to study the slope, aspect, and elevation of the terrain. This feature of AutoCAD Raster Design is also useful for conducting watershed studies, land suitability analysis, design, and management for a region. Figure 1-6 shows a digital elevation model of a region.

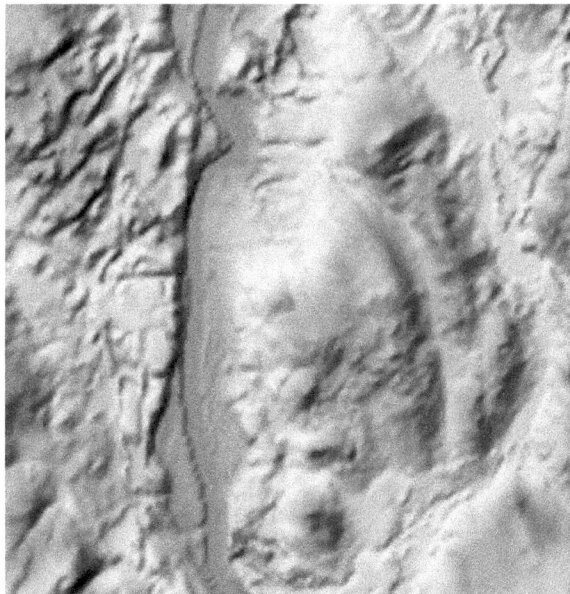

Figure 1-6 A digital elevation model of a region

Mechanical, Utility and Other Industries

In the past, all engineering products including the mechanical, civil, and other utilities were designed manually and were stored as engineering drawings on paper. Any modification in the existing design required engineers to undertake a careful study of these drawings making it a tedious and time consuming task.

Now-a-days, engineers use computers for planning and designing of almost every engineering system. For calculating the design parameters using computer, engineers are required to input data into the computer. Almost all designing software packages use input data in vector format. As a result, for the task of modifying the existing design, you are required to convert the existing paper drawings into vector data format. AutoCAD Raster Design provides a set of vectorization tools for easy and efficient conversion of the raster data to vector.

In addition, the REM tools in AutoCAD Raster Design allow you to edit raster entities. This software also provides the drawing cleanup tools that can be used to rectify any error that may be induced while scanning the paper copies. AutoCAD Raster Design is a useful tool for Town and Planning, City councils, and other similar departments which need to archive raster maps as historical datasets.

GETTING STARTED WITH AutoCAD RASTER DESIGN

As discussed earlier, AutoCAD Raster Design is an add-on product that can be installed on an AutoCAD based platform. On installing this product, the **Raster Tools** tab is added to the Ribbon interface of the AutoCAD based host application.

To initialize the AutoCAD Raster Design application, double-click on the desktop icon in Windows 8 operating system; the host application will start, refer to Figure 1-7. The ribbon interface of the host application will display the **Raster Tools** tab.

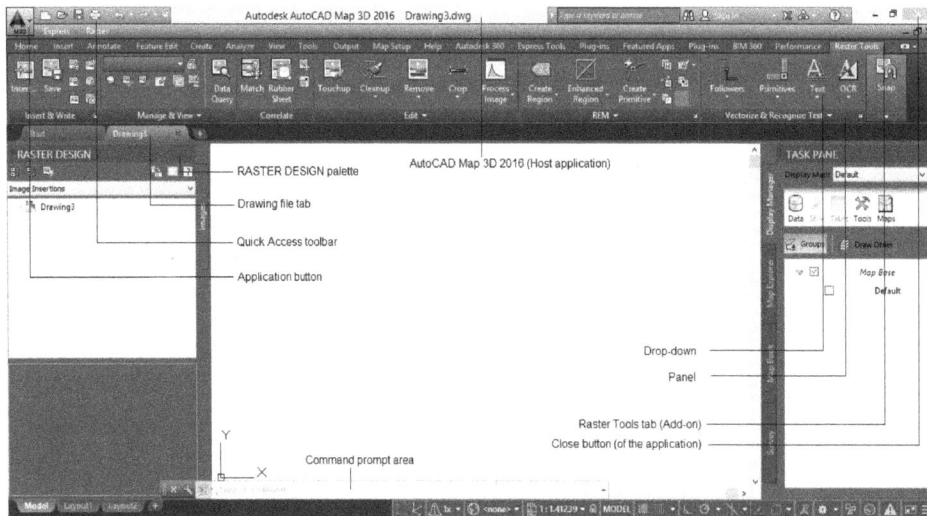

*Figure 1-7 AutoCAD Raster Design 2016 interface with the **Raster Tools** tab*

Note

For demonstration purpose, the author in this book has used AutoCAD Map 3D 2016 as the host application for installing AutoCAD Raster Design. The version of the host application and the add-on should be same.

Application Button

The Application button is displayed at the top-left corner of the interface. This button is used to display as well as close the Application Menu.

Application Menu

The Application Menu is displayed on choosing the Application button located at the top-left of the interface, refer to Figure 1-7. By default, the **Recent Documents** button is chosen in the Application Menu. As a result, the recently opened drawings will be listed on the right in the Application Menu. Click on the required file name in the list to open the file. To open a file that is not listed in this menu, choose the **Open** button in the Application Menu; the **Select File** dialog box will be displayed. Browse to the location of the required file. Click on the file name and then choose the **Open** button; the selected file will be opened in the drawing area. In AutoCAD Raster Design, you can specify the settings for the display, user preferences, files, and drafting parameters in the **Options** dialog box. To invoke this dialog box, choose the **Options** button displayed at the bottom-right of the Application Menu. Next, use the options in this dialog box to specify the required parameters. To exit AutoCAD Raster Design, choose the **Exit AutoCAD Map 3D 2016** button (in this case) from the Application Menu.

Ribbon

Ribbon interface in AutoCAD Raster Design contains tools organized into various tabs and panels based on their functionality, refer to Figure 1-8. When you start AutoCAD session for the first time, by default the Ribbon is displayed horizontally below the **Quick Access** toolbar. The Ribbon consists of various tabs. These tabs have different panels which in turn have tools arranged in rows. Some of the tools have a small black down arrow. This arrow indicates that there are other similar tools as well. To choose a tool, click on the down arrow; a drop-down will be displayed. Choose the required tool from the drop-down displayed. Note that when you choose a tool from the drop-down, the corresponding command will be invoked and the tool that you have chosen will be displayed in the panel.

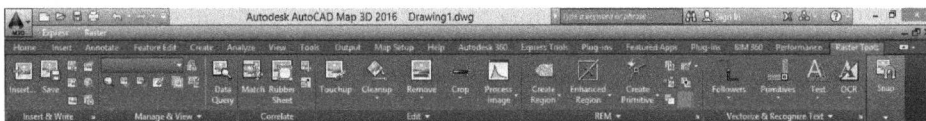

Figure 1-8 Ribbon in the interface

In this textbook, the tool selection sequence will be written as choose the **Circle** tool from **Raster Tools > REM > Create Primitive** drop-down. The tools which are not displayed within the available area of the panel are placed in the expandable area of the panel. Panels with an expandable area have a down arrow displayed to the right of its name. To view the tools in the expandable area of the panel, you can click on the down arrow. You can click on the push-pin in the expanded panel to keep it in the expanded state. Also, some of the panels have an inclined arrow at the lower-right corner. When you click on the inclined arrow, a dialog box is displayed. You can define the settings of the corresponding panel in this dialog box.

The base application allows you to change the default location of the Ribbon interface. To do so, right-click on the blank space in the Ribbon; a shortcut menu will be displayed. Next, choose the **Undock** option from this menu; the Ribbon is undocked. After undocking the Ribbon, you can move, resize, anchor, and turn on the auto-hide option for the display of the Ribbon. To do so, right-click on the heading strip in the Ribbon; a shortcut menu will be displayed. Choose the required option from this menu. For example, to vertically anchor the floating Ribbon to the left of the drawing area, right-click on the heading strip of the floating Ribbon, and choose the **Anchor Left <** option from the shortcut menu; the Ribbon will be anchored to the left.

You can also customize the display of tabs and panels in the Ribbon. To do so, right-click on any of the tools in it; a shortcut menu will be displayed. On moving the cursor over one of the options, a cascading menu will be displayed with a tick mark before all the options. Also, the corresponding tab or panel will be displayed in the Ribbon. Select/clear the desired option to display/hide a particular tab or panel. You can also reorder the display of panels in the tab. To do so, press and hold the left mouse button on the panel to be moved. Next, drag it to the required position and release the mouse button; the panel will be moved.

Drawing file Tabs

The drawing file tabs displayed above the drawing area, refer to Figure 1-7, show the drawings that are currently opened. Using these tabs, you can quickly switch between drawings. The order in which these tabs are displayed is dependent on the sequence in which the files were opened.

Drawing Area

The drawing area covers the major portion of the screen. In this area, you can draw objects by using various tools/commands. To draw an object, you need to define coordinate points. You can do so by using the pointing device. The cursor represents the position of the pointing device on the screen. There is a coordinate system icon at the lower-left corner of the drawing area.

To exit AutoCAD Raster Design, close the application; you will be prompted to save any unsaved work in your drawing. Choose the required action for saving the drawing file; the host application will be closed. The method of saving the image file is discussed in detail in further chapters.

Before you start using AutoCAD Raster Design, it is advised that you configure the general settings.

CONFIGURING OPTIONS IN AutoCAD RASTER DESIGN

You can configure AutoCAD Raster Design using the options in the **AutoCAD Raster Design Options** dialog box similar to customizing the program settings in the host software (AutoCAD platform). Configuring AutoCAD Raster Design, using various options this dialog box, is briefly discussed next.

AutoCAD Raster Design Options Dialog Box

To specify the settings in the AutoCAD Raster Design such as the default parameters for image insertion or settings for raster entity detection, choose the **Options-Image Default** button from the **Insert & Write** panel of the **Raster Tools** tab; the **AutoCAD Raster Design Options** dialog box will be displayed.

The **AutoCAD Raster Design Options** dialog box contains various tabs and options for configuring AutoCAD Raster Design. Different tabs in this dialog box are discussed next.

User Preferences Tab

The options in this tab are used to specify various global settings. These options are categorized into various areas, as shown in Figure 1-9.

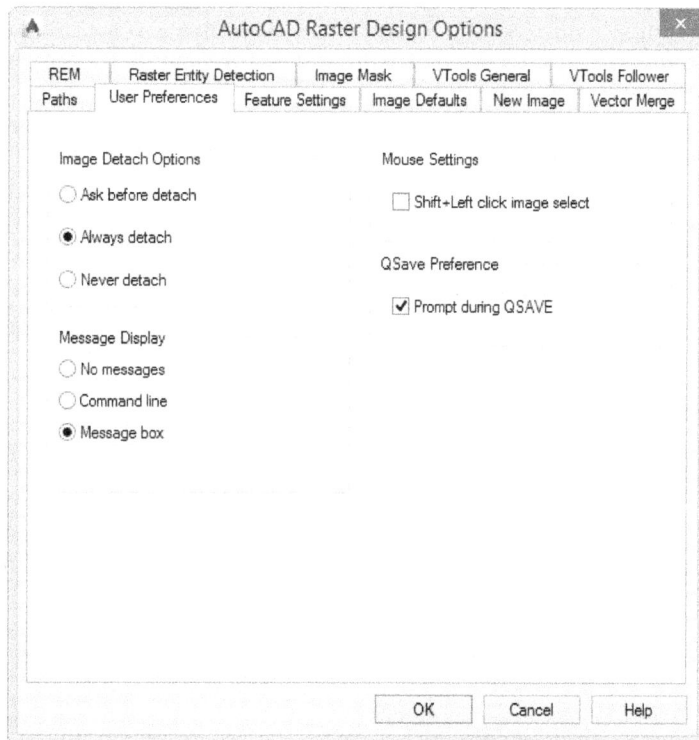

*Figure 1-9 The **User Preferences** tab of the **AutoCAD Raster Design Options** dialog box*

In the **Image Detach Options** area of this tab, you can specify the option for detaching the raster image by selecting the required radio button. Select the **Ask before detach** radio button if you want AutoCAD Raster Design to prompt you before detaching or retaining the image. Select the **Always detach** radio button, the image from the drawing will detach if its frame is erased. Select the **Never detach** radio button to retain the attached image after its frame is erased.

In the **Message Display** area of this tab, select the **No messages** radio button to restrict AutoCAD Raster Design from displaying messages. To display messages only in the Command prompt area, select the **Command line** radio button. Select the **Message box** radio button to display messages in a message box.

In the **Mouse Settings** area, select the **Shift+Left click image select** check box to select an image by clicking on it while holding the SHIFT key. This feature is useful when you zoom in the image and the image frame is not visible.

In the **QSave Preference** area, select the **Prompt during QSAVE** check box; AutoCAD Raster Design will prompt you to save images when you quick save the drawing file.

Paths Tab

Using the options in this tab, you can define settings for the image correlation files and the **AutoPaste** feature. This tab has two areas: **Correlation Search Paths** and **AutoPaste (ESP and GSX Compatibility)**. These areas are discussed next.

In the **Correlation Search Paths** area, you can specify the path of correlation source. Whenever you insert an image into the drawing, AutoCAD Raster Design will search the specified location for the correlation file. To specify the path, choose the **Browse** button corresponding to the **Read path** edit box; a menu will be displayed, as shown in Figure 1-10. To specify the location on your computer, choose the **Local** option from the menu displayed; the **Choose a Resource File Directory** dialog box will be displayed. In this dialog box, select the required folder containing the correlation file and choose the **Open** button; the dialog box will be closed and the path of the selected folder will be displayed in the **Read path** edit box. Alternatively, you can manually enter the location path of the correlation source in the **Read path** edit box. You can also specify a web address as correlation source location in the **Read path** edit box. Next, choose the **Web** option to specify a web address as correlation source location. To do so, browse to the location or specify the web address in the **Read path** edit box.

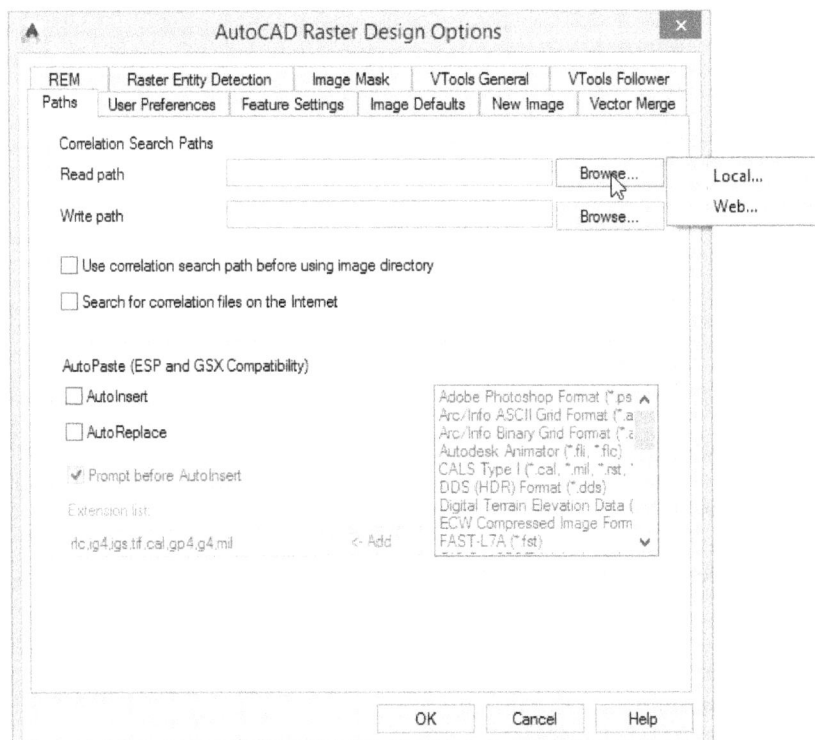

Figure 1-10 *The menu displayed on choosing the **Browse** button*

You can also specify the default path for saving or exporting correlation files. To do so, specify the location of the folder in the **Write path** edit box.

Next, select the **Use correlation search path before using image directory** check box to instruct AutoCAD Raster Design to search image correlation file from the specified default path before searching the image path stored in the drawing **Read path**. Similarly, you can select the **Search for correlation files on the Internet** to search for correlation sources on the internet.

The options in the **AutoPaste (ESP and GSX Compatibility)** area are cleared by default and used only when you are upgrading *.dwg* files using CAD Overlay ESP or GSX. Select the **AutoInsert** check box to automatically insert images into the drawing that has been created with ESP or GSX.

> **Note**
> *On selecting the **AutoInsert** check box, the **Prompt before AutoInsert** check box, the **Extension list** edit box, and the **Add** button will be activated.*

Next, you can select the **AutoReplace** check box to automatically search and insert images having same file name. AutoCAD Raster Design will insert new image using the insertion parameters of the old image.

Feature Settings Tab

You can use the options in this tab to create image thumbnails, specify the file locking settings, the width of the rub or crop line, and options for removing raster objects beneath the vector. Figure 1-11 shows various options in the **Feature Settings** tab.

To save the thumbnail of a raster image, select the **Save thumbnail with image** check box from the **Feature Settings** tab of the **AutoCAD Raster Design Options** dialog box. You can use the options in the **Locking Settings** area to lock the file you are editing. As the file is locked, it will deny write access to other users which are simultaneously using the same file. Selecting the **No locking** radio button will not lock the file and should be used only in the situation where you are sure that only one user will use the file. Selecting the **Lock file** radio button enables file locking when you are using a network that does not support DOS file sharing. And if your network supports DOS file sharing, select the **OS/Network locking** radio button to enable file locking.

> **Note**
> *The **Lock file directory** edit box will be activated on selecting the **Lock file** or **OS/Network locking** radio button.*

In the **Removal Method** area of the **Remove Underneath Settings** area, select the **REM** radio button to delete the raster object beneath the vector. To rub the raster below the vector, select the **Rub** radio button; AutoCAD Raster Design will rub the raster features along the vector as per the width specified in the **Rub/Crop Line Width** edit box. To specify the width of the rub or the crop line, specify the required value in the **Rub/Crop Line Width** text box. You can also specify the value in this edit box by picking two points from your drawing. To do so, choose the **Pick** button displayed at the right of the **Rub/Crop Line Width** edit box; the **AutoCAD Raster Design Options** dialog box will be closed. Next, specify two points in your drawing; the distance between the two specified points is displayed in the **Rub/Crop Line Width** text box.

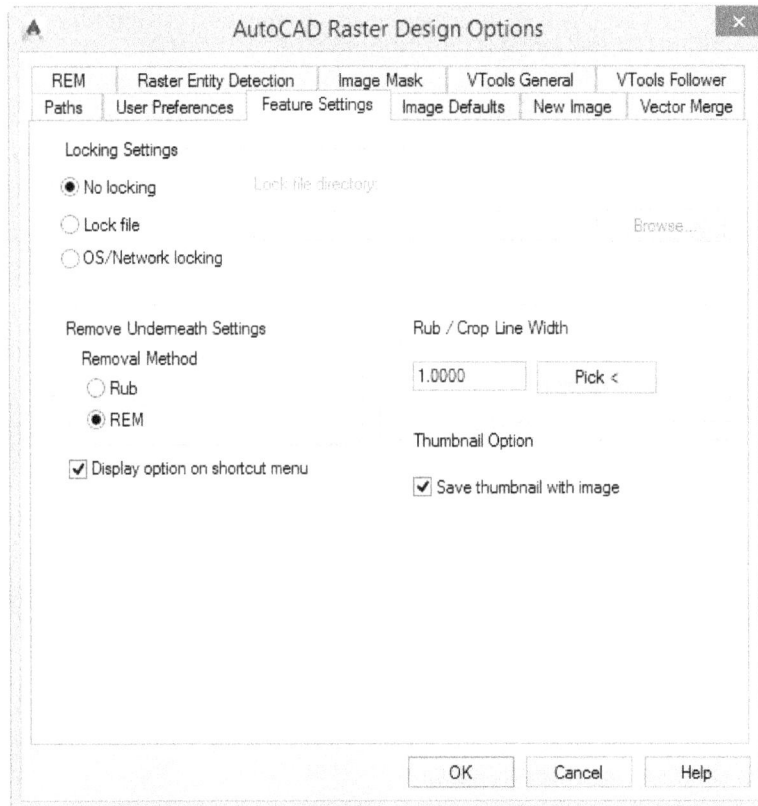

Figure 1-11 *The* ***Feature Settings*** *tab of the* ***AutoCAD Raster Design Options*** *dialog box*

Image Defaults Tab

In the **Image Defaults** tab of the **AutoCAD Raster Design Options** dialog box, you can set the default image correlation parameters. Figure 1-12 shows various options contained in this tab.

To set the default insertion point (lower left corner) of the raster, specify the values for the x, y, and z coordinates in the **X**, **Y**, and **Z** edit boxes, respectively. In the **Scale/Rotation** area, specify the scale factor and angle of rotation of the raster image in the **Scale** and **Rotation** edit boxes, respectively. Next, you will be required to specify the units for measuring in vertical axis. To do so, select the required unit from the **Vertical Units** drop-down list. To specify the density of pixel in the image, enter a value in the **Value** edit box in the **Density** area. Next, select the units for a density by selecting the required option from the **Units** drop-down list.

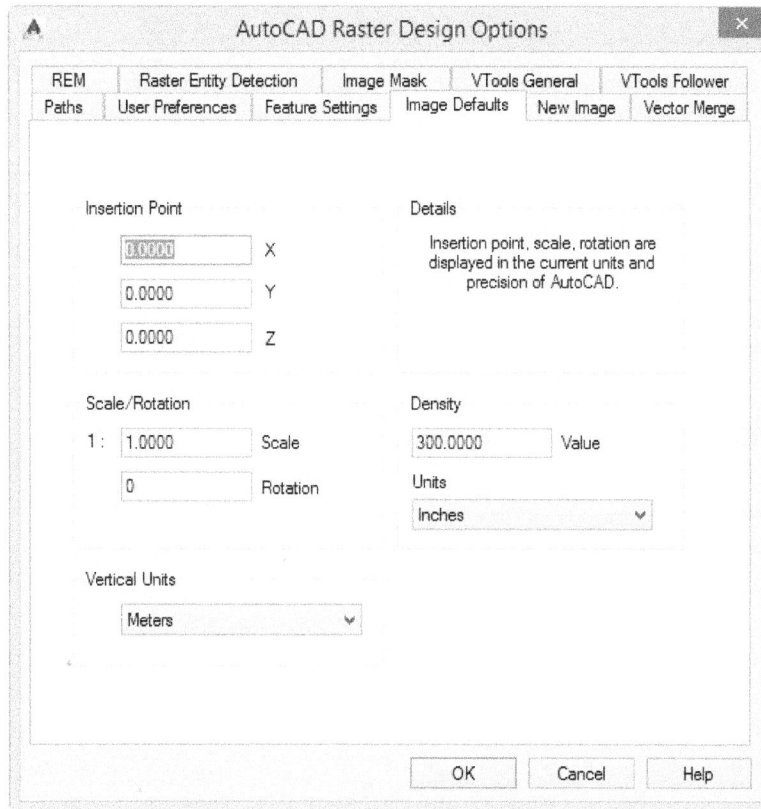

*Figure 1-12 Partial view of the **AutoCAD Raster Design Options** dialog box with the **Image Defaults** tab chosen*

New Image Tab

The options in the **New Image** tab are used to set the default parameters for creating a new image. You can specify the parameters of image size in the **Default Image Properties** area of this tab. Enter the width of the image in number of pixels in the **Width** edit box. You can also enter the required value in the Units edit box corresponding to the **Width** edit box, refer to Figure 1-13. Similarly, specify the height of the image in the **Height** edit box. Also, you can specify the image width in units which will be selected from the Units drop-down list corresponding to the **Density** edit box.

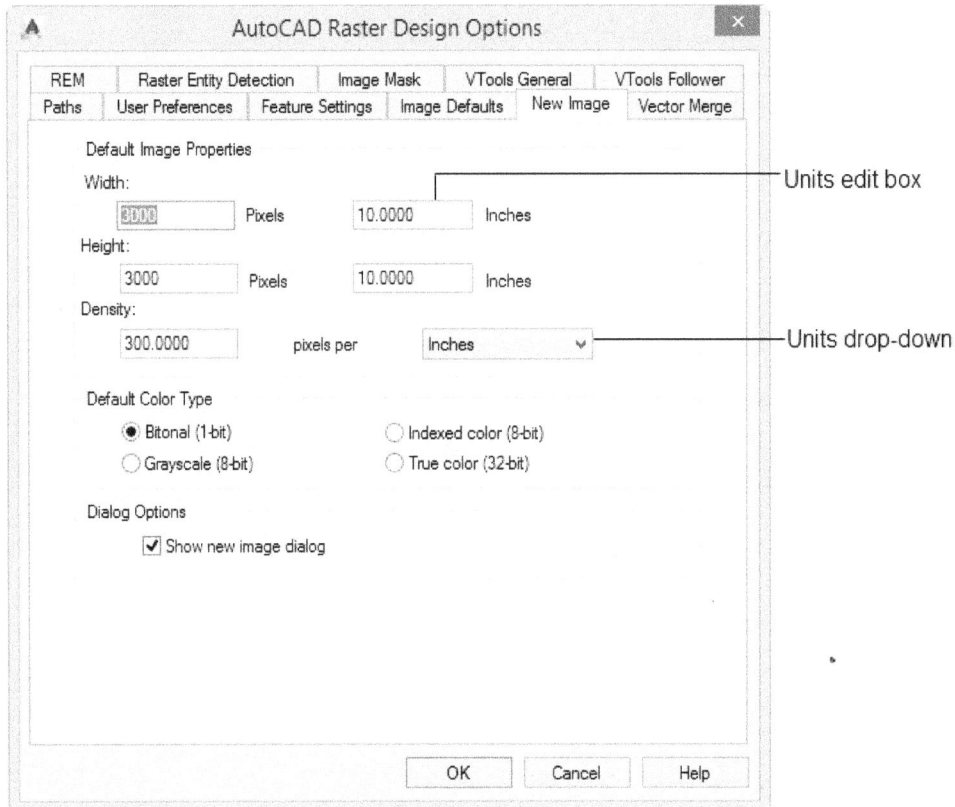

Figure 1-13 The New Image tab of the AutoCAD Raster Design Options dialog box

Note

1. On specifying the width of image in the Width edit box, AutoCAD Raster Design will calculate and display the width of the image in the Units edit box next to the Width edit box and vice-versa.

2. The value specified in the Density edit box should be the value at which your images are scanned.

In the **Default Color Type** area of this tab, assign default color type for the image by selecting the radio button corresponding to the required color type.

The **Show new image dialog** check box in the **Dialog Options** area is selected by default. On clearing this check box, the **New Image** dialog box will not be invoked on choosing the **New** button from the **Insert & Write** panel of the **Raster Tools** tab.

Image Mask Tab

Using the options in the **Image Mask** tab, you can set the properties for the image mask. Figure 1-14 shows the **Image Mask** tab of the **AutoCAD Raster Design Options** dialog box.

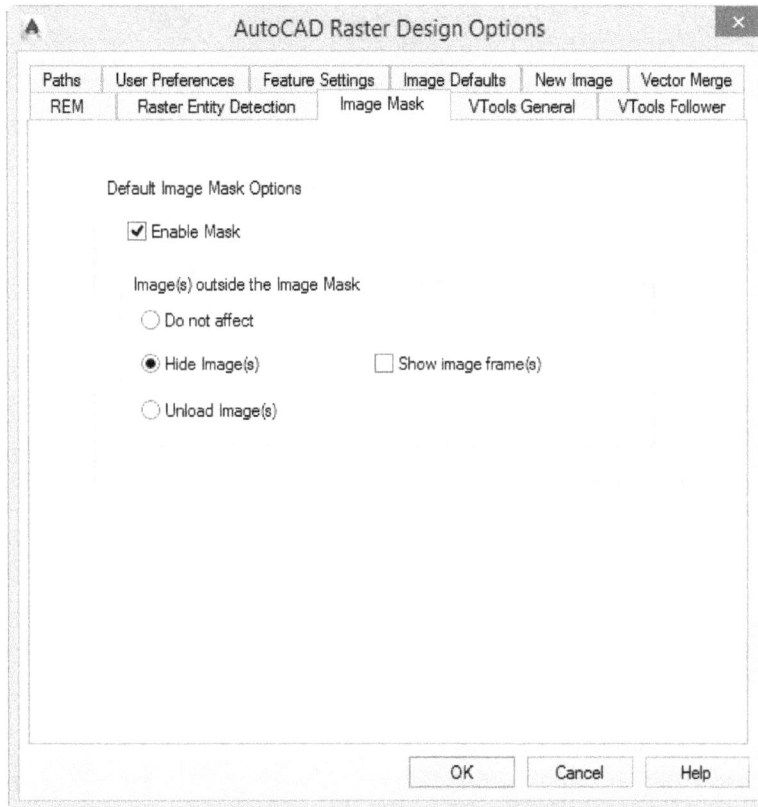

*Figure 1-14 The **Image Mask** tab of the **AutoCAD Raster Design Options** dialog box*

Select the **Enable Mask** check box from the **Default Image Mask Options** area to enable the existing mask in the drawing. Select the **Do not affect** radio button from the **Image(s) outside the Image Mask** area to ignore the effects of masking on the images that fall outside the mask boundary. To hide the images that fall outside the boundary of the mask, select the **Hide Image(s)** radio button. To unload images outside the mask boundary, select the **Unload Image(s)** radio button. To display the boundary of the hidden or unloaded images in the drawing, select the **Show image frame(s)** check box. Note that this check box will be disabled if you choose the **Do not affect** radio button.

REM Tab

Using the options in the **REM** tab, you can specify the color and the clipboard settings for the REM objects. Figure 1-15 shows the **REM** tab of the **AutoCAD Raster Design Options** dialog box.

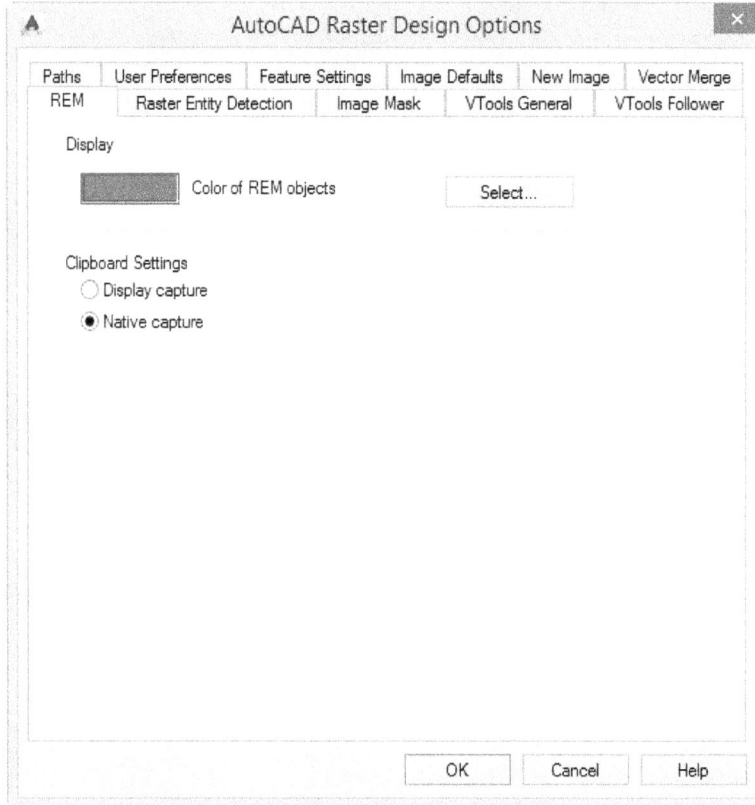

*Figure 1-15 The **REM** tab of the **AutoCAD Raster Design Options** dialog box*

By default, the REM objects are colored in red. To change the color of REM objects, choose the **Select** button in the **Display** area of the **REM** tab; the **Select Color** dialog box will be displayed. Select the required color in this dialog box and then choose the **OK** button; the **Select Color** dialog box will be closed and the selected color will be displayed in the **Display** area of the **REM** tab.

In the **Clipboard Settings** area of this tab, select the **Display capture** radio button to maintain the display scale and rotation of the REM objects that are copied to the clipboard using the **icopyss** command. To copy the REM object to the clipboard disregarding the display scale and rotation, select the **Native Capture** radio button.

Raster Entity Detection Tab

In the **Raster Entity Detection** tab, you can do various settings for detecting and following raster entities while working with the REM and vectorization tools. Figure 1-16 shows the options in this tab. These options are categorized into three areas that are discussed next.

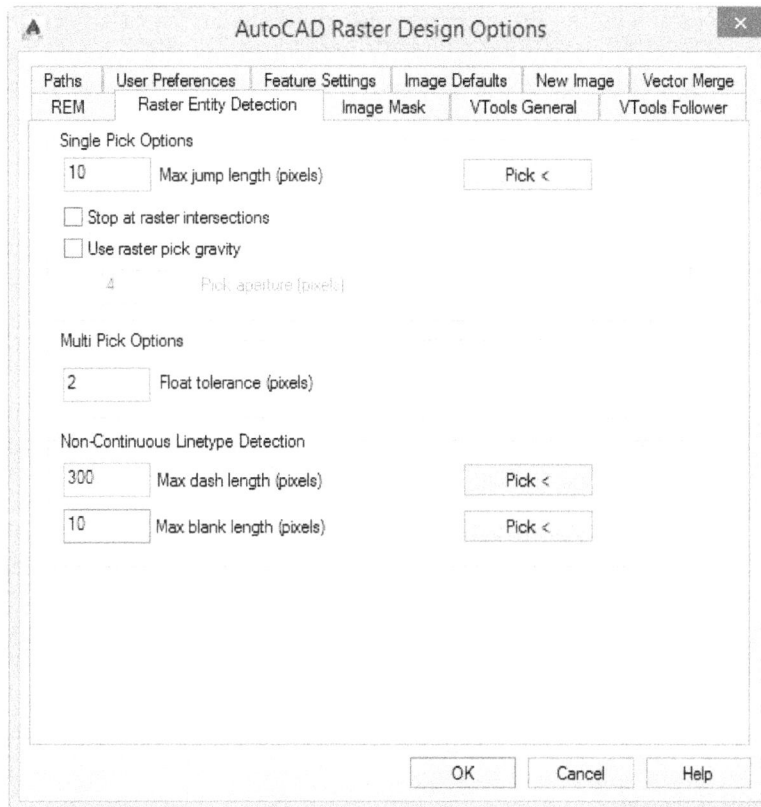

Figure 1-16 The Raster Entity Detection tab of the AutoCAD Raster Design Options dialog box

Single Pick Options Area

Using the options in this area, you can set the parameters for detecting raster entity in an image. Specify jump value in the **Max jump length (pixels)** edit box to ignore any discontinuity (break) in the given raster entity. While detecting a raster feature if AutoCAD Raster Design encounters a break in the feature it will calculate the length of the gap. If this length is less or equal to the value specified in the **Max jump length (pixels)** edit box, AutoCAD Raster Design will ignore this break and continue detecting the raster feature.

Specify the jump value by picking points in the drawing. To do so, choose the **Pick <** button from the **Single Pick Options** area; the **AutoCAD Raster Design Options** dialog box will be closed and you will be prompted to specify one end of the gap. Click at the required location to specify the point; you will be prompted to specify the other end. Click at the required location; the **AutoCAD Raster Design Options** dialog box will be displayed with the gap value specified in the **Max jump length (pixels)** edit box.

You can instruct AutoCAD Raster Design to stop detecting the raster feature beyond the points of intersections. To do so, select the **Stop at raster intersections** check box.

To select the raster feature within defined tolerance, select the **Use raster pick gravity** check box. Note that on selecting this check box the **Pick aperture (pixels)** edit box will be activated. You can specify the selection tolerance (radius of selection) in pixels in this edit box.

Multi Pick Options Area

In the **Float tolerance (pixels)** edit box of this area, you can enter the tolerance value in pixels. AutoCAD Raster Design will ensure that the selected raster entity is within the specified tolerance along the vector.

Non-Continuous Linetype Detection Area

By using the options in this area, you can specify the parameters for detecting a dashed line. To specify the maximum length of a dash, enter the value in the **Max dash length (pixels)** edit box. Alternatively, to specify the dash length graphically, choose the **Pick** button next to the **Max dash length (pixels)** edit box; the **AutoCAD Raster Design Options** dialog box will be closed and you will be prompted to specify one end of the dash. Click on the required location; you will be prompted to specify the other end of the dash. Click to specify the other end of the dash; the **AutoCAD Raster Design Options** dialog box will be displayed and the graphically specified value will be displayed in the **Max dash length (pixels)** edit box. Next, you can specify the maximum value of the gap between two consecutive dashes in the **Max blank length (pixels)** edit box. Alternatively, specify the gap value graphically as explained above.

VTools General Tab

Using the options in the **VTools General** tab, set the general behavior of the vectorization tools, as shown in Figure 1-17. The areas in this tab are discussed next.

Removal Method Area

In this area, select the required radio button to specify the method of removing the raster entity. To retain the raster feature, select the **None** radio button. To rub the raster, select the **Rub** radio button. On selecting this radio button, AutoCAD Raster Design will rub (erase) the underlying raster equivalent to the width specified in the **Rub/Crop Line Width** edit box of the **Feature Settings** tab. Select the **REM** radio button to delete the underlying raster entity.

To set the parameters for assigning layers and width to the vectors created using the vectorization tools, invoke the **Vector Separation Options** dialog box. To invoke this dialog box, choose the **Vector Separation** button in the **VTools General** tab, refer to Figure 1-17. The options in the **Vector Separation Options** dialog box are discussed further in the chapter.

Line, Circle, Arc and Polyline Settings Area

The options in this area are further categorized and displayed in various areas, refer to Figure 1-17. These areas are discussed next.

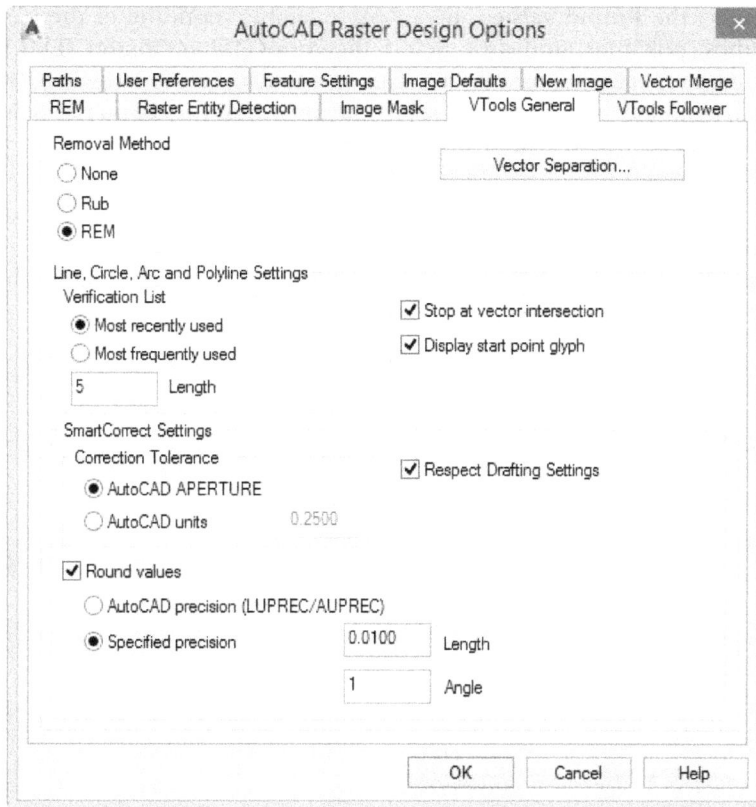

Figure 1-17 *The* **VTools General** *tab of the* **AutoCAD Raster Design Options** *dialog box*

In the **Verification List** area, select the required radio button to specify the type of value that will appear in the verification list. Next, specify the number of records to include in the verification list by entering a value in the **Length** edit box.

In this area, select the **Stop at vector intersection** check box to stop the vectorization tool, at vector intersection. To display a glyph at the start point, select the **Display start point glyph** check box.

In the **SmartCorrect Settings** area, you can select the required radio button from the **Correction Tolerance** area. Select the **AutoCAD APERTURE** radio button to set the allowable distance in pixels. Else, select the **AutoCAD units** radio button to specify a value in the edit box. You can also select the **Respect Drafting Settings** check box in this area to adhere to the specified drafting settings.

You can select the **Round values** check box to enable rounding of the length, angle, and radii of lines, polylines, and arcs. Select the **AutoCAD precision (LUPREC/AUPREC)** radio button to specify the precision, value that you set in the **Drawing Setup** dialog box. To customize the specified precision, select the **Specified Precision** radio button. On doing so, the **Length** and **Angle** edit boxes will be activated. Enter the required precision values in these edit boxes.

VTools Follower Tab

The options in this tab are used for vectorization by setting the parameters for the follower tools. Figure 1-18 shows various options in the **VTools Follower** tab.

*Figure 1-18 The **VTools Follower** tab of the **AutoCAD Raster Design Options** dialog box*

In this tab, the **Follower color** box displays the color of the vector. To change the color, choose the **Select** button; the **Select Color** dialog box will be displayed. Select the desired color in this dialog box and then choose the **OK** button; the **Select Color** dialog box will be closed and the selected color will be displayed in the **Follower color** box of the **Vtools Follower** tab. The **Pan to decision point** check box is selected by default. As a result, the drawing pans will automatically display the current decision point while using the **Followers** tools.

Next, select the **End current polyline if closed loop detected** check box for closing polylines and contour. On selecting this check box, the **Close tolerance in pixels** edit box will be activated.

Specify the value of tolerance in this edit box. AutoCAD Raster Design will close the polyline if it finds the start and end node of a polyline within tolerance specified in this edit box. You can also specify the value of tolerance graphically by choosing the **Pick** button, as explained in previous sections.

Select the **Post process points** check box to specify the degree of precision for drawing the lines and contours using the **Followers** tools. On selecting this check box, the **Polyline deviation** slider control will be activated. Setting the slider control to **Low** will create polylines with more vertices and thereby producing vectors that closely match the raster entity. Note that, a polyline feature with more vertices will require more storage space.

In the **Contour Settings** area of this tab, specify the settings for creating contours using the **Contour Follower** tool. To choose the object type of the contour, select an option from the **Contour creates** drop-down list. Select an option from the **Elevation** drop-down list to specify the method for specifying contour elevation. You can set the default elevation and the contour interval by specifying values in the **Preset elevation** and **Elevation interval** edit boxes, respectively. Note that the availability of these edit boxes will depend on the option selected from the **Elevation** drop-down list.

Note

*You can select the **Contour Object** option from the **Contour creates** drop-down list only if you are using AutoCAD Raster Design with AutoCAD Land Development Desktop.*

In the **3D Polyline Settings** area, enter the contour interval in the **Elevation interval** edit box. Next, select the **Use raster impact points only** check box if you want the **Followers** tools to pause only at locations where a polyline intersects a raster entity and to ignore other vertices. To prevent the **Followers** tools from pausing at raster intersection points due to the presence of speckles in the image, select the **Ignore raster speckles** check box; the **Speckle size (pixels)** edit box will be activated. Enter the speckle size in this edit box. You can also specify the speckle size on the screen by choosing the **Pick** button. On doing so, you will be prompted to pick a speckle or window from the bitonal image. After defining the window for the speckle size; the **AutoCAD Raster Design Options** dialog box will be displayed with the **Speckle size (pixels)** edit box displaying the specified speckle size. Note that the value of the speckle size should between 1 and 50 or else some necessary information might be deleted from the image considering it as speckle.

Vector Merge

The options in this tab are used to specify the settings for the vector merge operations. Figure 1-19 shows the **Vector Merge** tab in the **AutoCAD Raster Design Options** dialog box.

In the **Prompting Options** area of this dialog box, select the **Prompt to delete vector** check box. As a result, you will be prompted to delete all the selected vectors after the completion of the vector merge operation. If you do not want the prompt, ensure that this check box is cleared. When this check box is cleared, the radio buttons below this check box will be activated. Select the **Always delete vector** radio button to delete all the selected vectors after the vector merge operation. To retain all the selected vectors after the vector merge operation, select the **Never delete vector** radio button.

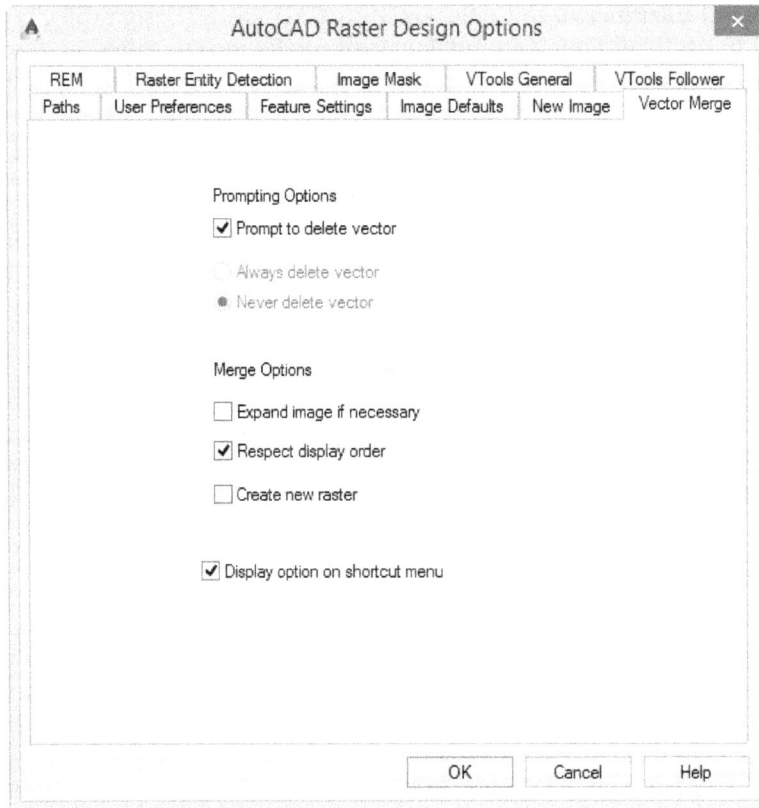

*Figure 1-19 The **Vector Merge** tab of the **AutoCAD Raster Design Options** dialog box*

In the **Merge Options** area of this tab, select the **Expand image if necessary** check box to expand the image to include the vector. To maintain the current display order on performing the vector merge, select the **Respect display order** check box. Next, select the **Create new raster** check box to create a new image rather than merging the vectors into the existing raster.

On specifying the required options, choose the **OK** button in the **AutoCAD Raster Design Options** dialog box; the dialog box will be closed and AutoCAD Raster Design will be configured as per the settings specified in this dialog box.

Vector Separation Options Dialog Box

The options in the **Vector Separation Options** dialog box are used to set the parameters such as line width and layers for the vectors features that are created using the vectorization tools (VTools). To invoke this dialog box, choose the **Vector Separation** button, refer to Figure 1-17, from the **VTools General** tab of the **AutoCAD Raster Design Options** dialog box. On doing so, the **Vector Separation Options** dialog box will be displayed, as shown in Figure 1-20.

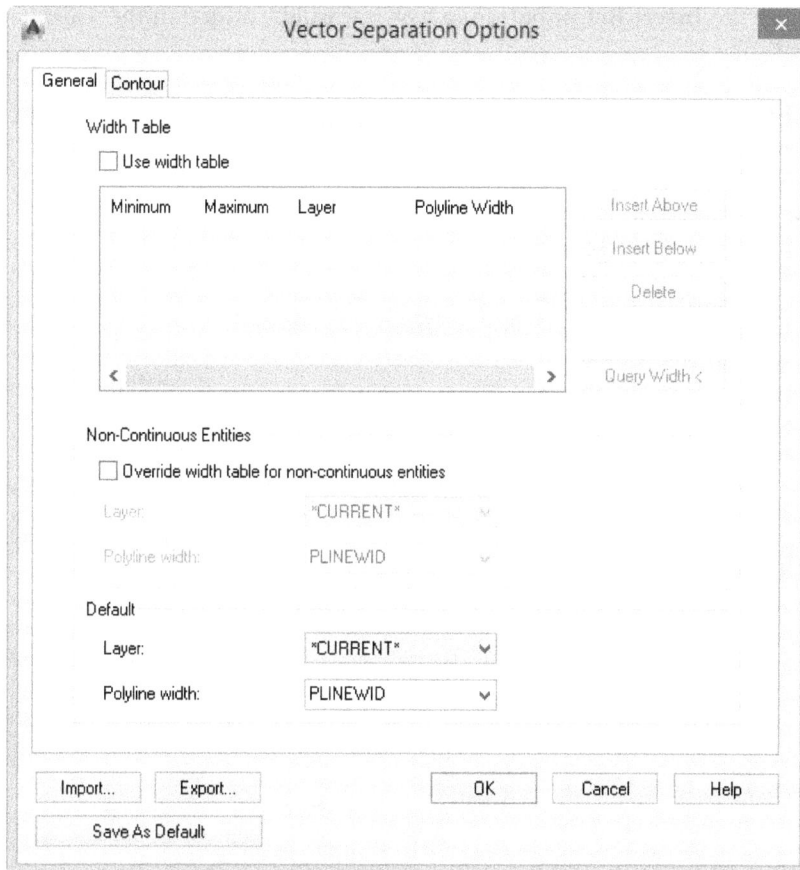

Figure 1-20 *The **Vector Separation Options** dialog box*

This dialog box contains two tabs. The options in these tabs are discussed next.

General Tab

Using the options in the **General** tab of the **Vector Separation Options** dialog box, specify the settings for assigning line width and layer to the vectors created using VTools. In this tab, you can also specify the method for handling noncontinuous (dashed) lines. The options in this tab are organized into three areas, refer to Figure 1-21. These areas in the **General** tab are discussed next.

Width Table Area

Select the **Use width table** check box to specify the layer of the created vectors based on the width of the raster feature from which they are created. The parameters for assigning layers to the vectors are defined in the table displayed in the **Width Table** area.

Note

*On selecting the **Use width table** check box, the various options in the **Width Table** area will be activated.*

Next, choose the **Insert Below** button; a new row will be added in the width table, refer to Figure 1-21.

*Figure 1-21 A new row added to the **Width Table** area*

To specify the minimum width of the raster entity in pixels, click in the **Minimum** cell of the created row; an edit box will be displayed. Specify the required value in the edit box and then press ENTER; the minimum width is set. Similarly, specify the maximum width in the **Maximum** cell.

If you are unsure of the width of the raster entity, you can use the query tool to find it. To do so, choose the **Query Width** button in the **Width Table** area; the dialog box will be closed and you will be prompted to choose the required raster entity in the drawing. Click on the raster entity. The **Entity Width Information** dialog box will be displayed, as shown in Figure 1-22. This dialog box shows the width of the selected raster entity in pixels and AutoCAD units.

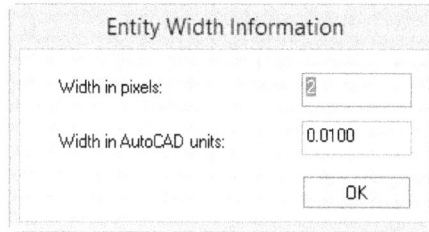

*Figure 1-22 The **Entity Width Information** dialog box*

Next, to specify the layer on which the vector is to be created, click on the **Layer** cell of the created row; a drop-down list will be displayed showing the layers in the drawing. Choose the required option from the drop-down list to set the layer. To set the width for the vector to be created, click in the **Polyline Width** cell of the row; a drop-down will be displayed, as shown in Figure 1-23.

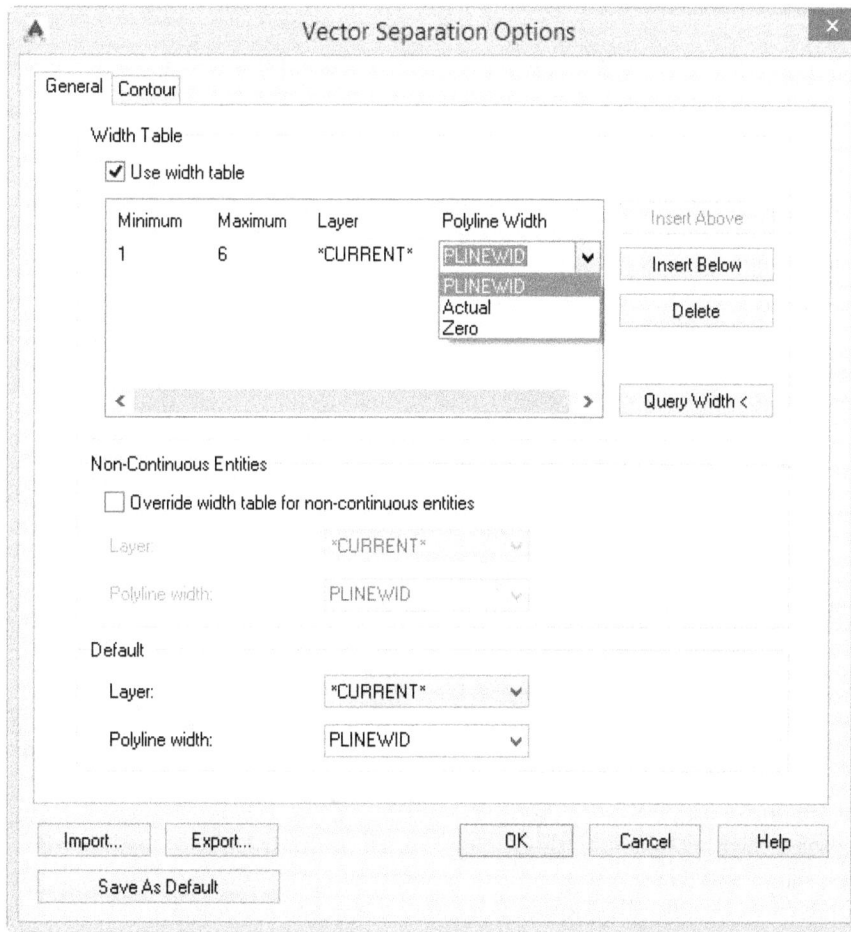

*Figure 1-23 The drop-down list displayed in the **Polyline Width** cell*

In this drop-down list, select the **Actual** option for creating vectors with line width equal to the width of the raster entity (from which it has been created). To create vectors with width equal to the line width defined in the layer property definition, select the **PLINEWID** option from the drop-down list. Else, select the **Zero** option to create vectors with zero width.

Non-Continuous Entities Area
In this area, you can override the width and layer settings specified in the **Width Table** area while creating vectors from the non-continuous lines. To do so, select the **Override width table for non-continuous entities** check box in this area; the **Layer** and **Polyline width** drop-down lists will be activated. Next, choose the required options from these drop-down lists to specify the override parameters.

Default Area
You can set the default width and layer parameters for the vector entities that do not fall in any defined range. To do so, select the required options from the **Layer** and **Polyline width** drop-down lists as explained earlier.

Contour Tab
The options in this tab are used to set the line width and layer parameters for the contours created using the VTools. The line width and the layers can be assigned based on the contour elevation interval. To do so, select the **Separate contours by elevation** check box, all the options in this tab will be activated, as shown in Figure 1-24.

In the **Minor** area, enter the required contour interval for minor contours in the **Interval** edit box. Next, to set the contour layer and width, select an option from the **Layer** and **Polyline width** drop-down lists, respectively. Similarly, specify the settings for the major contours in the **Major** area of this dialog box.

Tip
*The settings defined in AutoCAD Land Developer Desktop are used for specifying the layer and interval for the contours to be created using the VTools. To do so, select the **Use AutoCAD Land Development Desktop's settings** check box in the **Contour** tab of the **Vector Separation Options** dialog box.*

You can also specify the settings for the vector line width and layer in the **Vector Separation Options** dialog box by importing them from an external file. To do so, choose the **Import** button in this dialog box; the **Import Data** dialog box will be displayed. Next, in this dialog box select the file to be imported and then choose the **Open** button; the **Import Data** dialog box will be closed and vector settings will be imported in the **AutoCAD Raster Design Options** dialog box.

To export the line width and layer settings specified in the **Vector Separation Options** dialog box, choose the **Export** button in this dialog box; the **Export Data** dialog box will be displayed. Specify the name and location of the file to be exported in the **Export Data** dialog box and then choose the **Export** button; the **Export Data** dialog box will be closed and the settings specified in the **Vector Separation Options** dialog box will be saved to the file at the specified location.

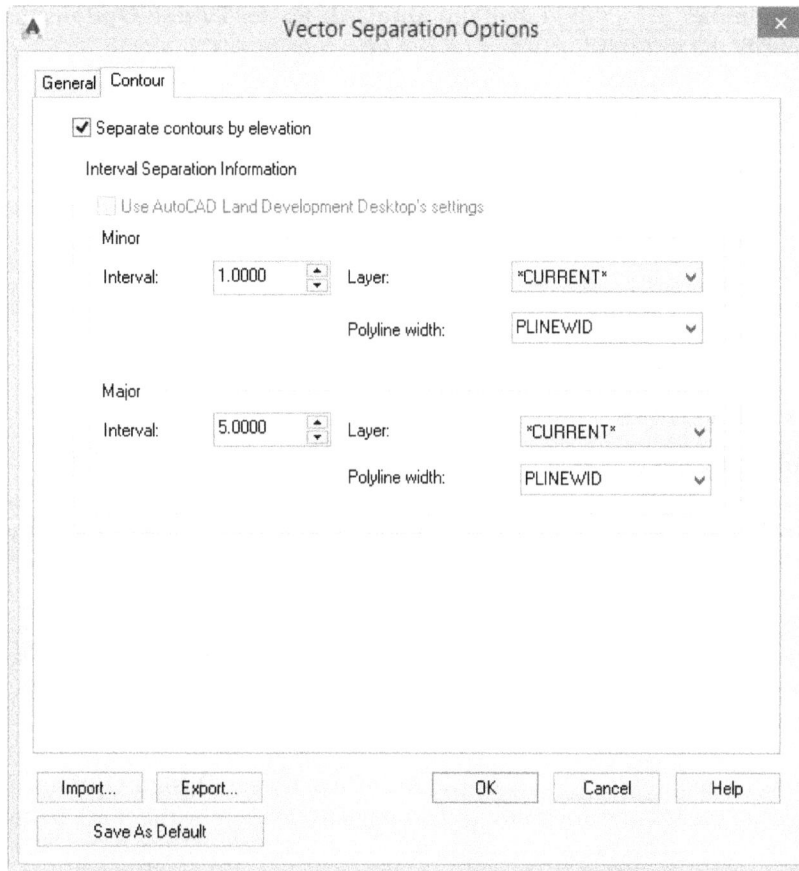

Figure 1-24 The Vector Separation Options dialog box

To make the current vector separation settings default, choose the **Save As Default** button. The current settings will be saved as the default settings. To close the **Vector Separation Options** dialog box without saving the settings, choose the **Cancel** button; the dialog box will be closed without applying the specified settings.

To do the required settings in the **Vector Separation Options** dialog box, choose the **OK** button; the vector separation settings will be set and the dialog box will be closed.

Self-Evaluation Test

Answer the following questions and then compare them to those given at the end of this chapter:

1. Which of the following tabs in the **AutoCAD Raster Design Options** dialog box is used to specify the settings for detecting raster entities while using the REM and vectorization tools?

    (a) **REM**   (b) **Raster Entity Detection**
    (c) **User Preferences**   (d) None of these

2. The options in the _____ tab of the **AutoCAD Raster Design Options** dialog box are used to specify the settings for detecting the non-continuous line type.

3. You can specify the options for detaching a raster image in the _____ tab of the **AutoCAD Raster Design Options** dialog box.

4. You can specify the vector separation settings in the **Vector Separation Options** dialog box by importing the settings file. (T/F)

5. The **Pick** button in the **AutoCAD Raster Design Options** dialog box is used to specify distance graphically. (T/F)

Review Questions

Answer the following questions:

1. You can use the _____ button in the **Vtools General** tab of the **AutoCAD Raster Design Options** dialog box to specify the values of polyline width for the vector layers created using the VTools.

2. You can use the options from the _____ in the **Vtools General** tab of the **AutoCAD Raster Design Options** dialog box to specify the entity type for the contour vectors created using the VTools.

3. The options in the _____ tab of the **AutoCAD Raster Design Options** dialog box is used to specify the settings for viewing the image as defined by the mask boundary.

4. The **Polyline deviation** slider is used to specify the degree of precision for drawing the lines and contours using the **Followers** tools. (T/F)

5. Selecting the **Lock file** radio button will enable to restrict the write access to other users that are simultaneously using the image file. (T/F)

Chapter 2

Insert, View, and Rubersheet Tools

Learning Objectives

After completing this chapter, you will be able to:
- *Insert an image into a drawing*
- *Transform the coordinate system of an image*
- *Correlate the images in a drawing*

INTRODUCTION

This chapter aims to introduce you to the interoperability feature of AutoCAD Raster Design. AutoCAD Raster Design tools enable you to easily insert, manage, and view scanned drawings, satellite images, aerial photographs, and digital elevation models. In this chapter, you will learn to insert a raster image into a drawing and you will also learn to transform the image coordinate system while inserting it into a drawing that has a different coordinate system.

In this chapter, we will also discuss some of the most commonly used raster operations such as scaling, moving, and rubbersheeting the inserted images. Moreover, you will learn to control the display of the inserted image using the options in the **RASTER DESIGN** palette.

INSERTING AN IMAGE

Sometimes we need to modify vector data or raster data. For that we need to insert these datasets into the drawing window of the software interface. AutoCAD Raster Design provides tools and options to insert raster images and drawing files in the drawing window using the **Insert** tool. The **Insert** tool allows you to specify the image correlation information to position, scale and rotate the image precisely so that every point on the image represents the true location on the surface of the earth. The image correlation parameters can be specified manually or by selecting the correlation source. Note that like the **Insert** tool, AutoCAD and AutoCAD MEP support the **ATTACH** command for inserting images. The procedure to insert a raster image using the **Insert** tool is discussed next.

Inserting a Raster Image Using the Insert Tool

Ribbon: Raster Tools > Insert & Write > Insert
Command: IINSERT

To insert a raster image into your drawing, choose the **Insert** tool from the **Insert & Write** panel of the **Raster Tools** tab; the **Insert Image** dialog box will be displayed, as shown in Figure 2-1. You can use the various options provided in the **Insert Image** dialog box to search, browse and select the raster image to be inserted into your drawing. The various options in this dialog box are discussed next.

In this dialog box, preview and the detailed information of the raster image are displayed on the right side. You can also browse for image data from the folder icons displayed on the left in this dialog box. Some of the options of this dialog box are discussed next.

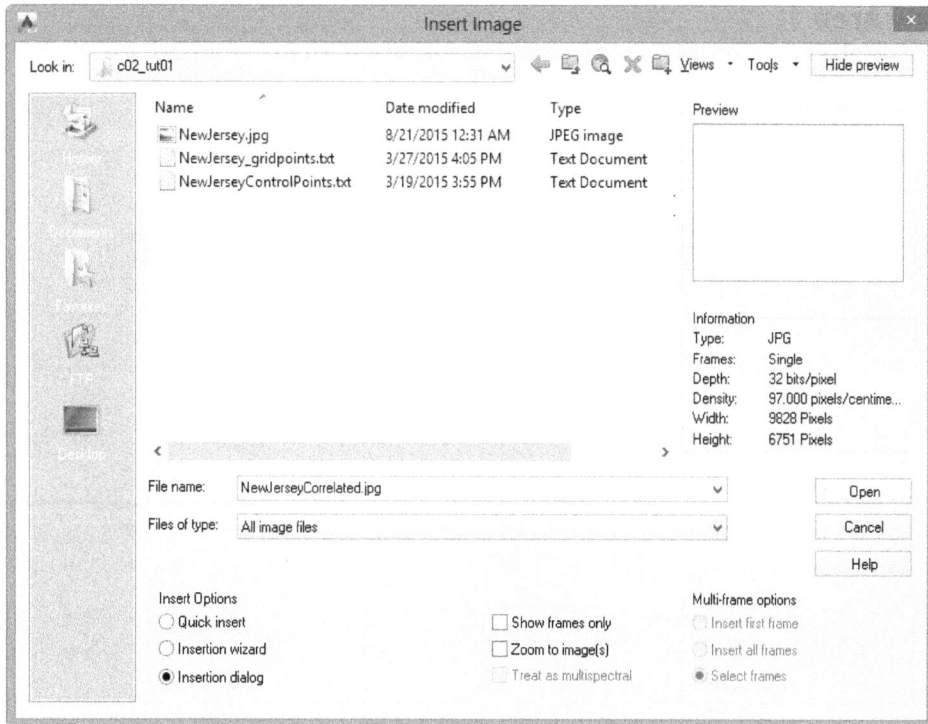

Figure 2-1 *The **Insert Image** dialog box*

Look in

You can choose an option from the **Look in** drop-down list to specify the folder or directory that contains the raster image. On selecting the required folder or directory, the content of the selected folder will be displayed in the list box below this drop-down list.

Search the Web

You can use the **Search the Web** button to find and select any raster image from the internet. On choosing this button, the **Browse the Web - Open** dialog box will be displayed. In this dialog box, specify the name and the URL of the file to be opened in the **Name or URL** edit box and then choose the **Open** button; the image from the specified path will be opened.

Files of type

This drop-down list contains the list of supported image file formats. You can choose an option from this drop-down list to filter the image format for selection.

> **Note**
> *AutoCAD Raster Design does not support .img file format, if you try to insert this file format, a message informing about an unknown format or invalid file will be displayed.*

File name

The **File name** drop-down list displays the name of the raster selected in this dialog box. This drop-down list also displays the path of recently used rasters. You can open a recently used raster by selecting an option from this drop-down list.

Preview Area

This area displays the preview of the selected image, as shown in Figure 2-2.

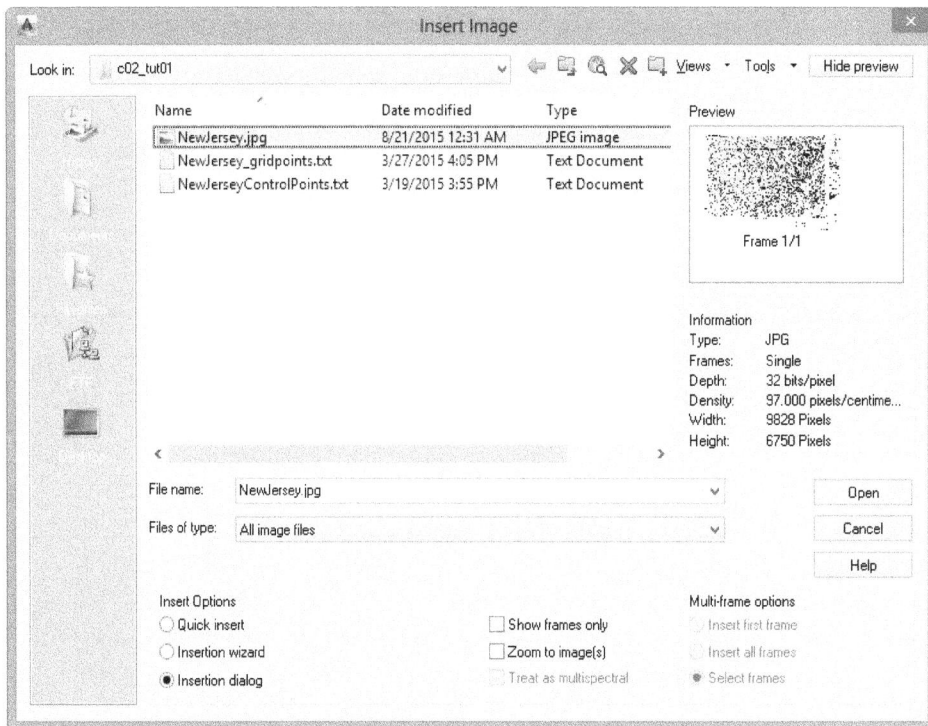

*Figure 2-2 The image preview displayed in the **Insert Image** dialog box*

Information Area

The **Information** area in this dialog box displays general information such as image type, size, density and so on of the selected raster image, refer to Figure 2-2.

Hide preview

You can use this button to toggle the display of the selected image in the **Preview** and **Information** area.

Insert Options Area

Using the options in this area, you can specify the method of inserting an image in the drawing. You can select one of the radio buttons such as **Quick insert**, **Insertion wizard**, or **Insertion dialog** from this area to provide correlation values to your drawing. You can also specify other insertion options by selecting the **Show frames only**, **Zoom to image(s)**, and **Treat as multispectral** check boxes to display image and zoom level.

The various image insertion options are discussed next.

Quick insert

You can select this radio button to insert an image into the drawing. While inserting the image using this option, AutoCAD Raster Design uses the correlation information available in the image, separate correlation file, or default values specified in the **AutoCAD Raster Design Options** dialog box.

Insertion wizard

Selecting this radio button will enable you to specify additional image insertion parameters like correlation source, positioning values, and so on. Select an image from the selection area of the **Insert Image** dialog box. Next, select the **Insertion wizard** radio button and then choose the **Open** button; the **Pick Correlation Source** page will be displayed, as shown in Figure 2-3.

In this page, you can specify the source for image correlation by selecting an option from the **Correlation source** drop-down list. If you select the **Image File** option from the **Correlation source** drop-down list, you will have to provide correlation values in the **Correlation Values** area. If you select the **Default Values** option from the **Correlation source** drop-down list, the image will be inserted using default values provided in the image.

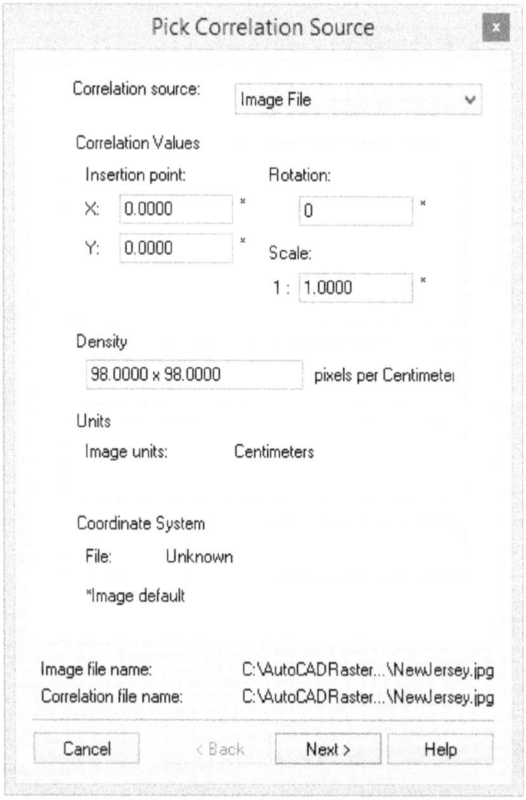

*Figure 2-3 The **Pick Correlation Source** page*

In the **Correlation Values** area, the **X** and **Y** display boxes display the coordinate values for the insertion of the raster image. Also, the rotation and scale values are displayed in their corresponding display boxes. The values displayed in the **Correlation Values** area depends upon the option selected from the **Correlation source** drop-down list. Note that, if the inserted image is already rectified, then the rectified X and Y coordinate values will be displayed in the **X** and **Y** display boxes.

The value in the **Density** display box displays the pixel density of the raster image. The image unit is displayed in the **Units** area below the **Density** display box, refer to Figure 2-3. The **Coordinate System** area displays the Coordinate Reference System (CRS), if applied, of the selected raster image.

On specifying the required options in this page, choose the **Next** button; the **Modify Correlation Values** page will be displayed, as shown in Figure 2-4. By default, the options in the **Modify Correlation Values** page will display the correlation values specified in the **Pick Correlation Source** page.

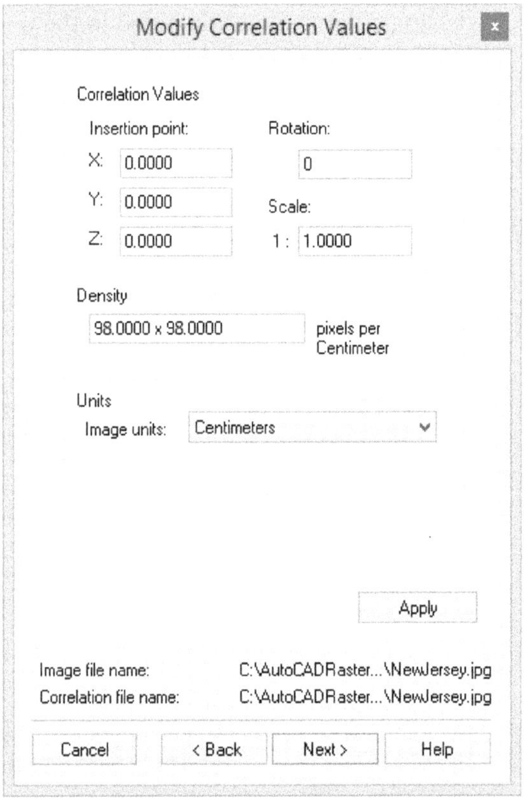

Figure 2-4 The Modify Correlation Values page

You can use the options in this page to modify the correlation parameters for the image to be inserted.

Note

*The values can be edited only in the **Modify Correlation Values** dialog box.*

To modify a correlation parameter, specify the required value in the **Correlation Values** area. For example, to modify the different image unit, select an option from the **Image units** drop-down list. To modify the scale of the image, specify the required value in the **Scale** edit box. Note that to enlarge the image, specify a scale value larger than 1 in the **Scale** edit box and to reduce the size of the image, specify a scale value between 0 and 1.

On modifying the required correlation parameters, choose the **Apply** button; the modified correlation values will be applied to the image. Choose the **Next** button; the **Insertion** page will be displayed, as shown in Figure 2-5.

Figure 2-5 *The **Insertion** page*

Using the options in the **Insertion** page, you can graphically or numerically specify the correlation parameters for inserting an image.

To specify the parameters numerically, enter the values in the required areas as explained in the previous sections. Alternatively, to graphically specify the values, choose the **Pick** button in the **Correlation Values** area; the **Insertion** page will be closed and you will be prompted to specify the base point for inserting the image. Click on the required location in the drawing; you will be prompted to specify the angle of rotation of the image. Note that moving the cursor in the drawing will rotate the image frame attached to the cursor about the specified base (insertion) point. Rotate the frame to the required angle by moving the cursor and then click to specify the rotation angle; you will be prompted to specify the corner. Note that the frame of the image is now fixed on the basis of the specified insertion point and rotation angle. You will also notice that the size of the frame changes on moving the cursor. Move the cursor to choose the required frame size and then click; the **Insertion** page will be displayed. This page will now show the correlation parameters that were specified graphically. You can specify the color of the image frame. To do so, choose the **Select** button in the **Color** area of the page; the **Select Color** dialog box will be displayed, as shown in Figure 2-6. Choose the required color in this dialog box and then choose the **OK** button; the dialog box will be closed and the selected color will be displayed in the color preview box next to the **Select** button in the **Color** area of the **Insertion** page. Choose the **Apply** button in the wizard to apply the specified parameters.

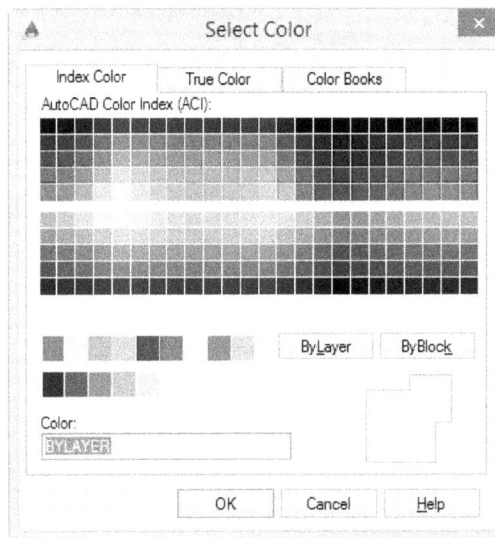

Figure 2-6 The Select Color dialog box

Next, choose the **Finish** button in the **Insertion** wizard; the image is inserted in the drawing using the correlation parameters specified in the wizard.

Insertion dialog

This radio button will enable you to specify additional image insertion parameters. To insert an image; select an image from the selection area of the **Insert Image** dialog box. Next, select the **Insertion dialog** radio button and then choose the **Open** button; the **Image Insertion** dialog box will be displayed with the **Source** tab chosen, as shown in Figure 2-7.

The options in this dialog box are similar to the **Pick Correlation Source** page, as discussed earlier.

Note

*In case you are using a GIS application like AutoCAD Map 3D as the base software for installing AutoCAD Raster Design, then while inserting an image having coordinate system different than the coordinate system assigned to the drawing, the **Transform** page will be displayed after the **Modify Correlation Values** page. You can use the options in this page to transform the coordinate system of the image.*

Show frames only

To display only the frame of the inserted image in the drawing, select the **Show frames only** check box in the **Insert Image** dialog box, refer to Figure 2-2.

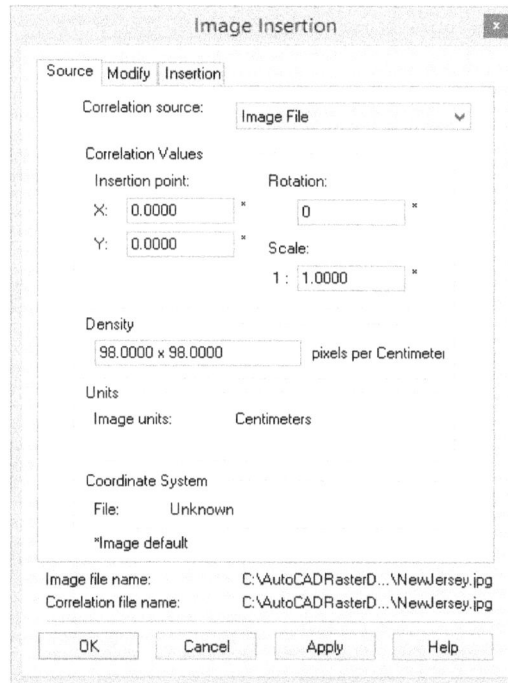

*Figure 2-7 The **Source** tab of the **Image Insertion** dialog box*

Zoom to image(s)

You can select this check box to automatically zoom to the extents of the inserted image. This check box is cleared by default as a result AutoCAD Raster design will not zoom to the extents of the inserted image.

Treat as multispectral

This option will be activated only when two or more images are selected in the selection area of the **Insert Image** dialog box. Using this option you can insert the selected images as a multispectral dataset. Inserting a multispectral image into the drawing is discussed in detail further in this book.

Multi-frame options Area

You can select an option in this area to specify which image frame you want to insert from a multiframe image. A multiframe image dataset consists of multiple images within a single file. Note that, each image within the multiframe file has a separate frame.

To activate this area, select a multiframe image like *.ntf* extension file in the selection area of the **Insert Image** dialog box. If you select the **Insert first frame** radio button, the first frame from the multiframe dataset will be displayed in the drawing. Similarly, if you select the **Insert all frames** radio button, it will insert all the image frames of the multiframe dataset.

Note
*If the **Quick insert** radio button is selected in the **Insert Image** dialog box, AutoCAD Raster Design will directly insert the selected image into the drawing without prompting for any additional input parameters from the user.*

In case you are using AutoCAD Raster Design with a GIS application such as AutoCAD Map 3D, AutoCAD Raster Design provides the options for coordinate transformation while inserting an image having coordinate system different than that of the drawing.

TRANSFORMING IMAGE COORDINATE SYSTEM

As discussed earlier, while inserting an image having coordinate system different than that of the drawing environment, AutoCAD Raster Design will display the **Transform** page after the **Modify Correlation Values** page, refer to Figure 2-8.

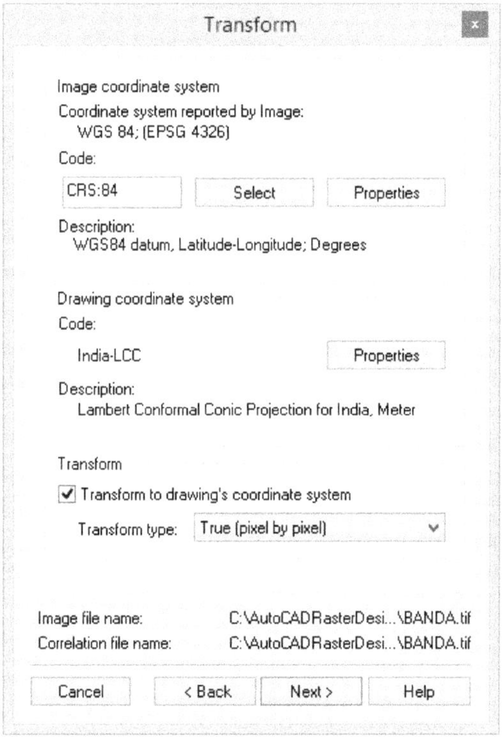

*Figure 2-8 The **Transform** page*

On selecting the **Insertion dialog** radio button in the **Insert Image** dialog box while inserting an image in the drawing, the **Transform** tab will be displayed in the **Image Insertion** dialog box. Note that, before inserting the desired image, you need to assign coordinate system to the drawing environment, only then this tab will be displayed.

You can use the options in the **Transform** tab for transforming the image coordinate system. Note that, to activate the options in this tab you need to select the **Transform to drawing's coordinate system** check box in the **Transform** area. The options in the **Transform** tab are discussed next.

Tip
If the image coordinate system and drawing coordinate system have different coordinate values, the image will not open in your drawing area.

Image coordinate system Area

In most cases, AutoCAD Raster Design will identify the coordinate system of the selected image and will display it in the **Image coordinate system** area of the **Transform** page.

You can also specify a coordinate system for the image using the options in this area. To do so, enter the coordinate system code in the **Code** edit box. AutoCAD Raster Design will search the coordinate system library for the specified code. Alternatively, you can specify the coordinate system by selecting the required coordinate system from the library. To do so, choose the **Select** button; the **Select Coordinate System** dialog box will be displayed, as shown in Figure 2-9.

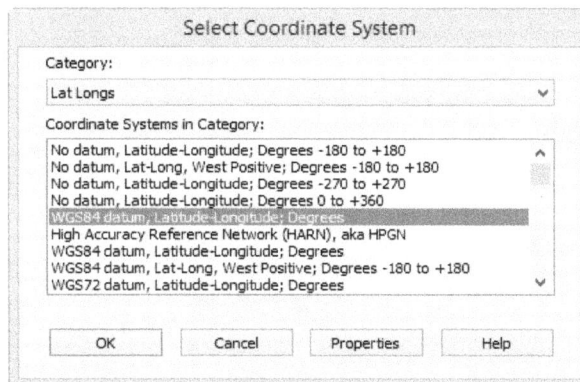

Figure 2-9 The Select Coordinate System dialog box

In the **Select Coordinate System** dialog box, select the required category of the coordinate system by selecting an option from the **Category** drop-down list. On doing so, the coordinate systems in the selected category will be displayed in the **Coordinate Systems in Category** list box. Next, choose the required coordinate system in the **Coordinate Systems in Category** list box, refer to Figure 2-9, and then choose the **OK** button; the code of the selected coordinate system will be displayed in the **Code** edit box of the **Transform** page.

In AutoCAD Raster Design, you can view the properties of the selected coordinate system. To do so, choose the **Properties** button in the **Image coordinate system** area; the **Coordinate System Property** dialog box will be displayed, as shown in Figure 2-10. Note that you cannot edit the CRS parameters using the options in the **Coordinate System Property** dialog box.

Drawing coordinate system Area

The **Drawing coordinate system** area of the **Transform** page displays the coordinate system of the drawing environment. To view the properties of the drawing coordinate system, choose the **Properties** button in this area; the **Coordinate System Property** dialog box will be displayed, refer to Figure 2-10.

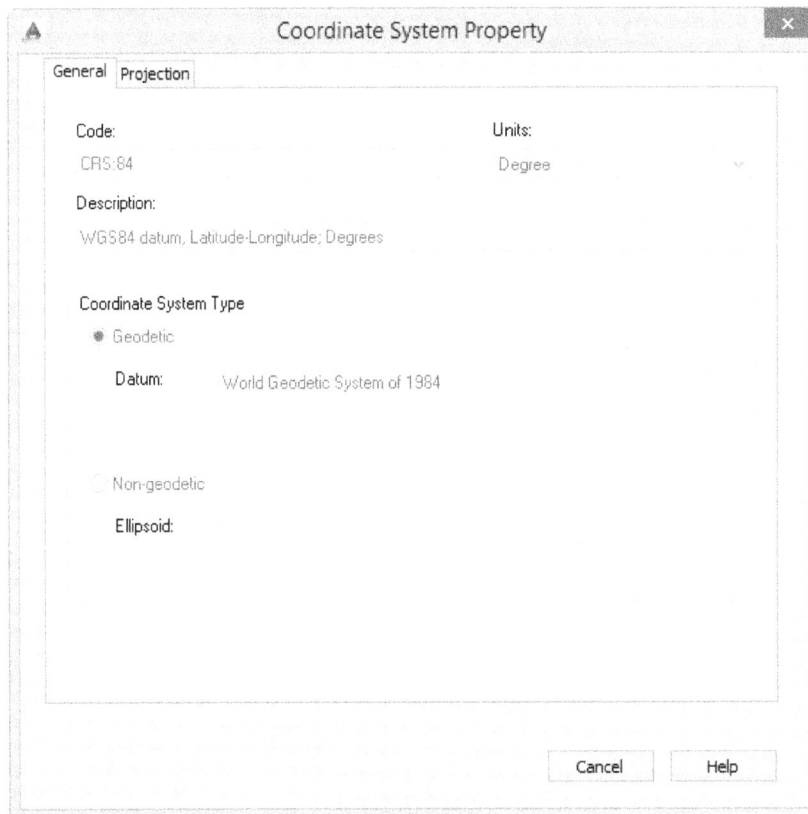

*Figure 2-10 The **Coordinate System Property** dialog box*

Transform Area

Select the **Transform to drawing's coordinate system** check box in the **Transform** area to enable the transformation of the image to the drawing coordinate system. Note that, all the options in the **Transform** page will be deactivated if you clear this check box.

You can choose the type of transformation by selecting an option from the **Transform type** drop-down list. The various options in this drop-down list are discussed next.

True (pixel by pixel)

The **True (pixel by pixel)** transformation type is applied to every pixel in the image. After applying this transformation type, the result will be more accurate. However, this is a slow process and requires large amount of resources for computing transformation.

4 point rubbersheet

You can select this option to enable image transformation using 4 points on the image. This transformation produces the most accurate results.

3 point affine

You can choose this option to enable affine transformation of the image. An affine transformation can differentially scale, skew, rotate, and translate a raster image. You need atleast 3 control points on the image to perform this method.

On specifying the required options in the **Transform** page, choose the **Next** button; the **Insertion** page will be displayed.

Note

*If you select the **Transform to drawing's coordinate system** check box in the **Transform** page, the options in the **Correlation Value** area of the **Insertion** page will be disabled.*

In the **Insertion** page, specify the required options as explained earlier in the chapter and then choose the **Finish** button; the selected image will be inserted in the drawing area.

The image inserted in the drawing is also added into the Tree View of the **RASTER DESIGN** palette. You can view the inserted image by clicking on the tree view symbol in the **RASTER DESIGN** palette. You can use this palette to view the properties of the inserted images. The various options in this palette can also be used to manage the raster images in the drawing. The various components of the **RASTER DESIGN** palette are discussed next.

RASTER DESIGN PALETTE

Ribbon:	Raster Tools > Manage & View > Manage
Command:	IMANAGE

The **RASTER DESIGN** palette in AutoCAD Raster Design is used to manage the inserted rasters in your drawings. Using this palette you can control the color and visibility of the image. You can also use this palette to view the image properties and related data.

To invoke this palette, choose the **Manage** tool from the **Manage & View** panel of the **Raster Tools** tab; the **RASTER DESIGN** palette will be displayed, as shown in Figure 2-11.

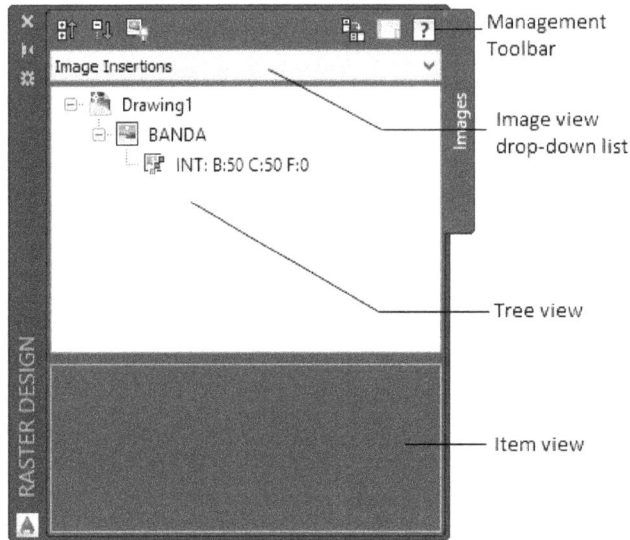

Figure 2-11 *The* ***RASTER DESIGN*** *palette*

The options and areas in the **RASTER DESIGN** palette are discussed next.

Management Toolbar

You can select the **Image Insertions** option from the Image view drop-down list. On doing so, the **Image Insertions** view will be displayed in the **RASTER DESIGN** palette. You can also display the **Image Data** view in the **RASTER DESIGN** palette by choosing the **Image Data** option from the Image view drop-down list.

The Management toolbar in the **RASTER DESIGN** palette contains buttons that can be used to control the display of the items. You can use the **Expand Tree** button to display the hierarchy of objects in the Tree view. You can use the **Collapse Tree** button to hide the hierarchy in the Tree view.

You can either display the image preview or the data related to the image as a table in the Item view of the **RASTER DESIGN** palette. To do so, choose the **Show Image Preview** button in the Management toolbar.

Note that on right clicking in the Item view area of the **RASTER DESIGN** palette, a shortcut menu will be displayed. The options displayed in this shortcut menu will depend on the option selected such as **Image Insertions** / **Image Data** in the Image view drop-down list. You can use the options in the shortcut menu to manage the images in the drawing.

Tree View

The tree view in the **RASTER DESIGN** palette displays the list of images in the drawing in a hierarchical view. Figure 2-12 shows the Tree view in the **RASTER DESIGN** palette in the **Image Insertions** view.

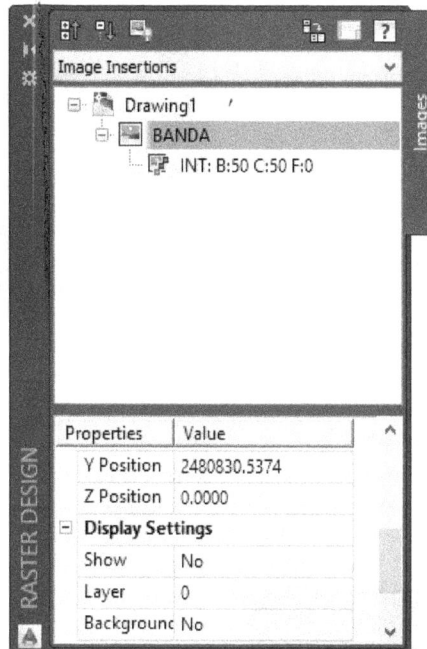

*Figure 2-12 The Tree view in the **RASTER DESIGN** palette*

Item View

The Item view in the **RASTER DESIGN** palette displays the properties of the selected image in a table. You can define the display and order of the columns in the table. To do so, right click on the column name in the Item view; a shortcut menu will be displayed, as shown in Figure 2-13. Next, choose any option from the shortcut menu. On doing so, the columns related to the selected option will not be displayed in the Item view area.

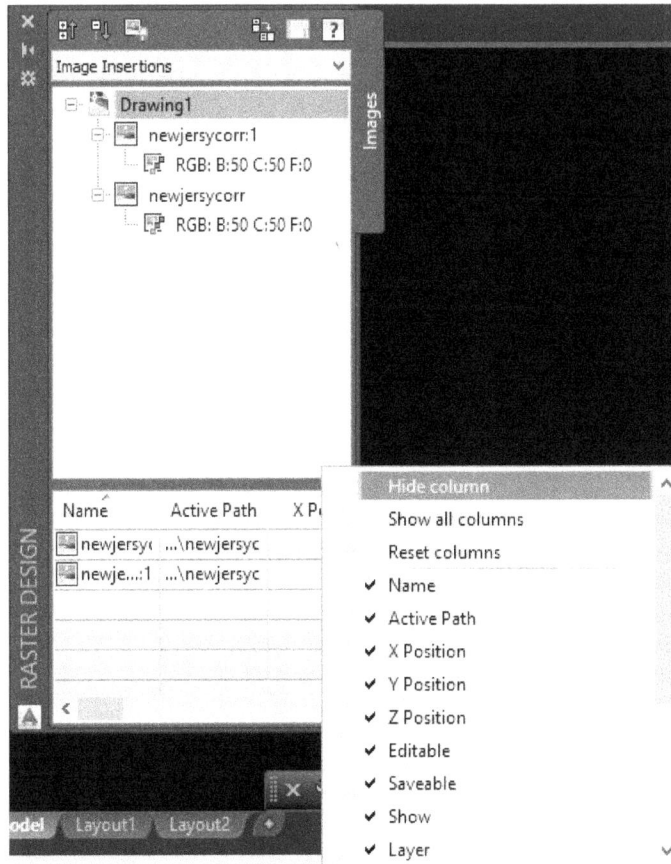

Figure 2-13 *Shortcut menu displayed on right clicking in the Item view of the **RASTER DESIGN** palette*

You can display the preview of the selected image in the Item view of the **RASTER DESIGN** palette. To do so, ensure that the **Image Insertions** option is selected in the Image view drop-down list. Next, choose the required image from the Tree view and then choose the **Show Image Preview** button from the Management toolbar; the preview of the selected image in the Tree view will be displayed in the Item view, as shown in Figure 2-14.

Note
*You can toggle between the table and image view in the Item view of the **RASTER DESIGN** palette only if you have selected the **Image Insertions** view option from the Image view drop-down list in the Management toolbar.*

As discussed earlier, you can manage the inserted images in the drawing using the options in the **Image Insertions** view and **Image Data** view of the **RASTER DESIGN** palette. These options are discussed next.

Image Insertions View

To display the **Image Insertions** view, choose the **Image Insertions** option from the Image view drop-down list in the **RASTER DESIGN** palette. In this view, you can display the table or image preview in the Item view as discussed earlier.

You can also use this view for managing image insertions and their color maps. To do so, expand the drawing name node in the **RASTER DESIGN** palette; the tree view will display the list of images inserted in the current drawing, refer Figure 2-14. Next, right click on the required image name; a shortcut menu will be displayed, as shown in Figure 2-15. The various options in this shortcut menu are discussed next.

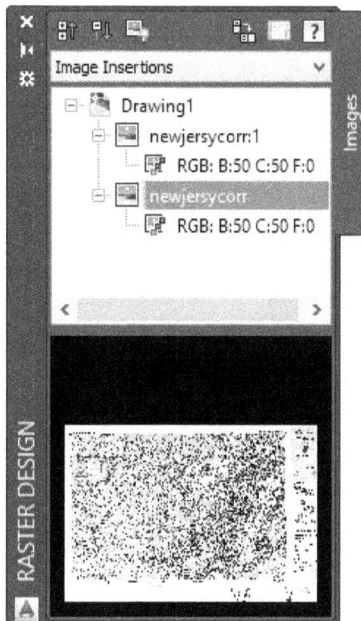

Figure 2-14 The Item view in the **RASTER DESIGN** palette displaying the preview of the selected image

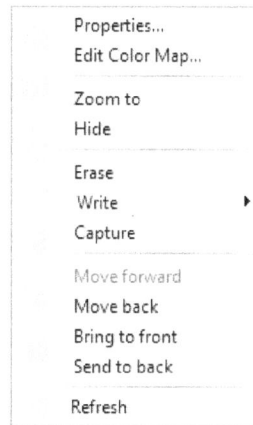

Figure 2-15 The shortcut menu displayed in the **Image Insertions** view of the **RASTER DESIGN** palette

Properties Option

You can choose the **Properties** option from the shortcut menu to display the properties of the selected image in the **PROPERTIES** palette. Figure 2-16 displays the **PROPERTIES** palette showing properties of a raster image.

Edit Color Map Option

You can choose the **Edit Color Map** option from the shortcut menu to edit the color map for the selected image. To do so, ensure that the **Image Insertions** option is selected in the Image view drop-down list of the **RASTER DESIGN** palette. Next, right-click on the required image; a shortcut menu will be displayed, refer to Figure 2-15. Choose the **Edit Color Map** option from the shortcut menu; the **Image Adjust** dialog box will be displayed, as shown in Figure 2-17. You can specify various image parameters using the options in this dialog box.

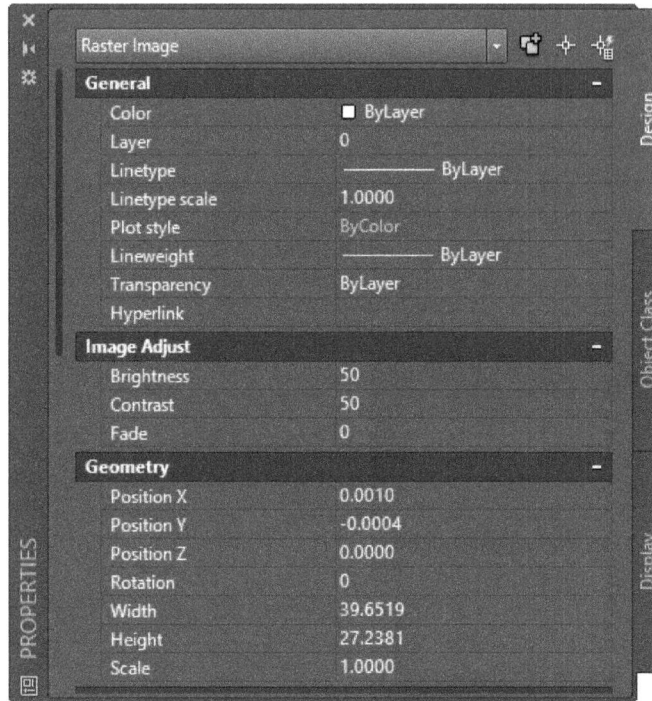

Figure 2-16 *The **PROPERTIES** palette*

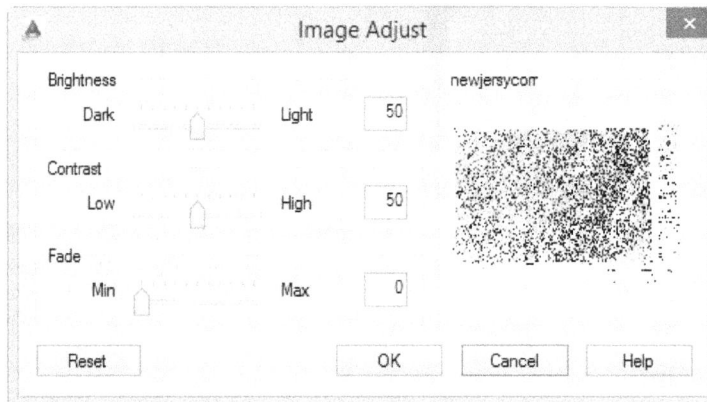

Figure 2-17 *The **Image Adjust** dialog box*

Image Adjust dialog box

Using the options in this dialog box, you can adjust the brightness, contrast, and fade values of the inserted image. You can adjust the image brightness by moving the **Brightness** slider, refer to Figure 2-17. Moving the slider toward the **Dark** end will reduce the brightness. To increase the brightness, move the slider toward the **Light** end of the slider. You can also specify the brightness value by entering the required value in the corresponding edit box. Similarly, you can adjust the **Contrast** and **Fade** values of the image by using the **Contrast** and **Fade** slider in the dialog box.

On specifying the required parameters, choose the **OK** button; the specified adjustments will be applied to the image in the drawing. To reset the image adjustment parameters to the default settings, choose the **Reset** button in the **Image Adjust** dialog box; the values will be reset to default.

You can also edit the color of a bitonal image. To do so, insert a bitonal image in the drawing. Choose the **Edit Color Map** option from the shortcut menu; the **Select Color** dialog box will be displayed, as shown in Figure 2-18. You can set the color in the image using the options in this dialog box.

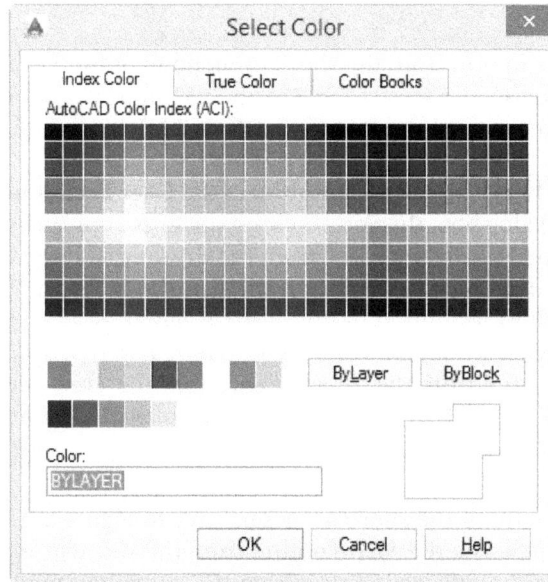

Figure 2-18 *The Select Color dialog box*

Select Color dialog box

In this dialog box, you can specify the required color by selecting it from the **AutoCAD Color Index (ACI)** palette in the **Index Color** tab, refer to Figure 2-18.

In the **True Color** tab of this dialog box, you can also manually define and assign a color using the **Hue Saturation Luminance** (**HSL**) or the **Red Green Blue** (**RGB**) color model.

In the **Color Books** tab of this dialog box, you can specify color using third-party color books or user-defined color books.

Select the required color from the **Select Color** dialog box and then choose the **OK** button; the dialog box will be closed and the selected color will be assigned to the bitonal raster.

Zoom to Option

You can choose the **Zoom to** option from the shortcut menu to zoom to the extents of the selected image.

Hide/Show

You can choose the **Hide** option from the shortcut menu to switch off the display of the selected image in the drawing. On choosing this option the selected image will be hidden in the drawing. To display a hidden image, invoke the shortcut menu and then choose the **Show** option from it; the selected image will display in the drawing.

> **Note**
> *Hiding an image will not detach (remove) the raster image from the drawing. The **Hide** option simply turns off the display of the image in the drawing.*

Erase Option

You can choose the **Erase** option from the shortcut menu to erase the selected image from the current drawing. On doing so, the selected image and its frame will be deleted from the current drawing and from the Tree view of the **RASTER DESIGN** palette.

If you have selected the **Always detach** radio button in the **User Preferences** tab of the **AutoCAD Raster Design Options** dialog box, then on choosing the **Erase** option in the shortcut menu, the erased image will be automatically detached from the drawing. On the other hand, if you have selected the **Ask before detach** radio button in the **AutoCAD Raster Design Options** dialog box then on choosing the **Erase** option in the shortcut menu, you will be prompted to specify whether you want to detach the erased image. If you choose not to detach the erased image, then selected image will be deleted from the drawing and from the Tree view of the **Raster Design** palette, but the Tree view of the **Image Data** view will display an unreferenced data definition.

Write Option

You can use the **Write** option in the shortcut menu, refer to Figure 2-15, to save or export the selected image. To do so, choose the **Write** option from the shortcut menu; a cascading menu will be displayed. Choose the required option from this cascading menu to save or export the selected image.

Capture Option

If you are unable to save or export an image, you can capture a snapshot of the image. To do so, choose the **Capture** option from the shortcut menu; the AutoCAD Raster Design will capture the snapshot of the selected image and will also insert it into the current drawing as a new insertion. Now you can save or export this snapshot as a new image. Note that, you can also capture the snapshot of the selected image using the **Capture** tool from the **Insert & Write** panel in the **Raster Tools** tab.

Image Display Order Options

The shortcut menu also provides various display order options such as **Move forward**, **Move back**, **Bring to front**, and **Send to back**, refer to Figure 2-15. You can use these options to manage the display order of the selected image.

Image Data

The **Image Data** view, refer to Figure 2-19, displays the data-centric view of the rasters in the **RASTER DESIGN** palette. You can invoke the **Image Data** view by selecting the **Image Data** option from the Image view drop-down list in the **RASTER DESIGN** palette. This view is best suited for working with image data definition and creating new image insertions. The Tree view for the **Image Data** view displays various objects and their symbols which are as follows:

Table 2-1 *Symbols and their description in the **Image Data** view*

	Image data definition
	Band group
	Band group metadata
	Data band
	Color map
	Image insertion

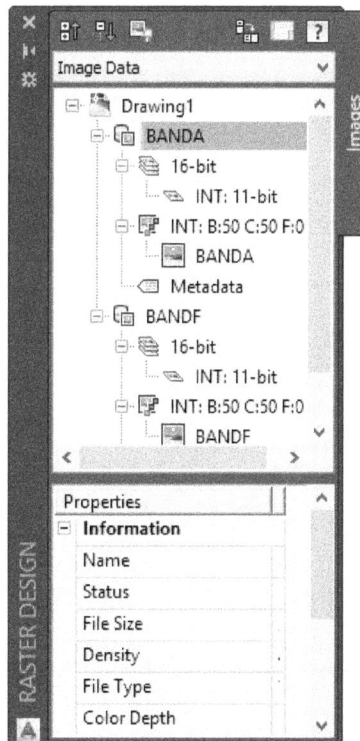

Figure 2-19 *The **Image Data** view in the **RASTER DESIGN** palette*

To perform an action such as creating a new image insertion or detaching an image in the **Image Data** view, right click on the required image in the Tree view; a shortcut menu will be displayed, as shown in Figure 2-20. Next, you can select the required option from the shortcut menu. The options in this shortcut menu are discussed next.

Properties
New Insertion
Detach
Reload
Unload
Embed
Un-embed
Browse path...
Save path
Clear path
Refresh

Figure 2-20 *The shortcut menu displayed in the* **Image Data** *view of the* **RASTER DESIGN** *palette*

Properties Option

You can choose the **Properties** option from the shortcut menu to invoke the **Data Definition** dialog box. This dialog box displays the data definition information for the selected image insertion, refer to Figure 2-21.

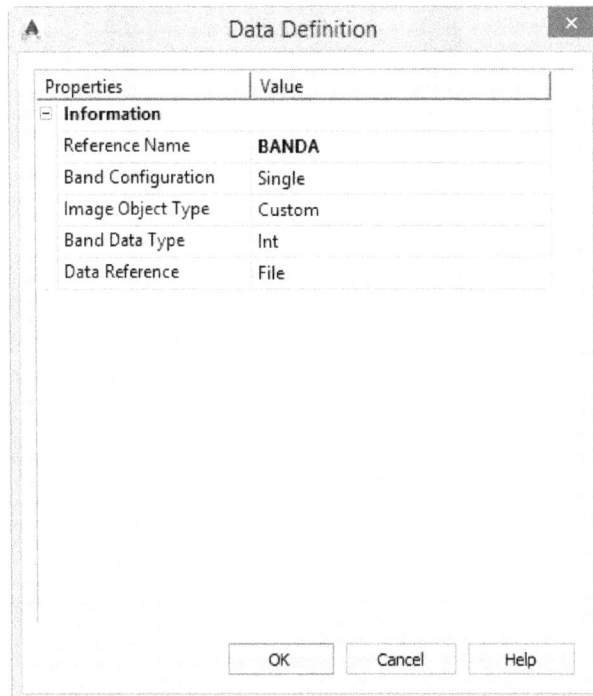

Properties	Value
⊟ **Information**	
Reference Name	**BANDA**
Band Configuration	Single
Image Object Type	Custom
Band Data Type	Int
Data Reference	File

Figure 2-21 *The* **Data Definition** *dialog box*

New Insertion Option

You can choose this option from the shortcut menu to create a new insertion for the selected raster image. On choosing this option, based on the type of raster selected in the Tree view, AutoCAD Raster Design will directly create a new insertion or will prompt you to specify additional parameters for creating the insertion. The new insertion is also displayed as a node in the Tree view of the **RASTER DESIGN** palette.

Detach Option

You can completely remove the image along with its data definition from the current drawing by choosing the **Detach** option from the shortcut menu. On doing so, the selected image is completely removed from the drawing and also from the **RASTER DESIGN** palette.

Unload/Reload Options

You can choose the **Unload** option from the shortcut menu to remove the selected image from the display. Unloading an image will remove the image from the display. To reload the unloaded image, you can choose the **Reload** option from the shortcut menu.

Browse path Option

You can invoke the **Choose a New Directory** dialog box by choosing this option from the shortcut menu. Using the options in this dialog box you can navigate to a different source file and open it.

Save path Option

Using this option, you can save the path of the file.

Clear path Option

You can choose the **Clear path** option from the shortcut menu to delete the saved path from the selected data definition.

CORRELATING INSERTED IMAGES

Images inserted without specifying correlation parameters may result in placing of those images at improper location or even with incorrect scale. Sometimes, images such as scanned maps or aerial photos are distorted which may lead to incorrect results. The use of raster images inserted at the correct location and scale is more important in applications such as Geographical Information Systems (GIS), as it requires georeferenced data for performing spatial analysis. It is therefore required that you perform image rectification so that the image is placed at the required location with the correct scale value and is free of any distortion. This can be achieved by performing various raster operations such as move, scale, and rotate. AutoCAD Raster Design also provides the **Rubber Sheet** tool using which you can georeference a raster image file for further use.

Correlating raster images can be a simple process using a single image correlation operation such as scaling or moving raster image or it can be done using a complex procedure that requires the use of multiple image correlating operations. The various methods in AutoCAD Raster Design used for correlating images are discussed next.

Displacing Image

Ribbon: Raster Tools > Correlate > Displace
Command: IDISPLACE

In situations where raster image is not inserted at a required location in the drawing, you can use the **Displace** tool to move and adjust image to the required location. Note that, using the **Displace** tool will not alter the scale and rotation of the image in the drawing.

To move an image in the drawing, choose the **Displace** tool from the **Correlate** panel of the **Raster Tools** tab; you will be prompted to specify the base point for the raster image to be moved. Click at the required location to specify the base point for the image; you will be prompted to specify the destination point for moving the raster image. Click on the drawing to specify the displacement point; the selected raster image will be moved to the specified point. Figure 2-22 shows the schematic representation of moving the inserting image using the **Displace** tool.

Note
If a drawing contains more than one raster image, AutoCAD Raster Design will prompt you to select the raster to be moved, scaled, and matched.

Scaling an Image

Ribbon: Raster Tools > Correlate > Scale
Command: ISCALE

You can use the **Scale** tool to change the scale of the inserted image. You can set the scale of an image to your desired scale value. Note that using the **Scale** tool will not change the rotation of the image in the drawing.

To change the scale of the selected image, choose the **Scale** tool from the **Correlate** panel in the **Raster Tools** tab; you will be prompted to specify the base point of the image. Click to specify base point. Next, you will be prompted to specify the source distance. Specify the source distance by clicking on the image and then press enter. Alternatively, you can specify the source distance by clicking two points in the drawing. On specifying the source distance you will be prompted to specify the destination distance. Specify the destination distance at the Command prompt area; the image will be scaled as per the specified parameters, refer to Figure 2-23.

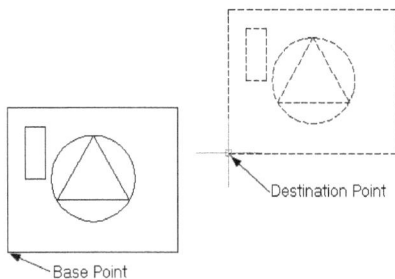

Figure 2-22 Moving the raster image

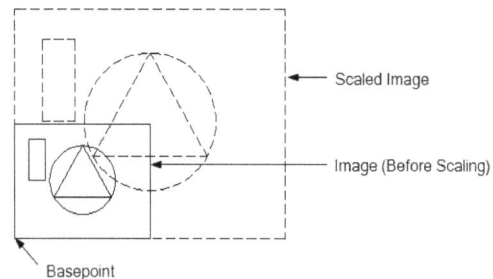

Figure 2-23 Scaling the raster image

Matching Image Points

Ribbon: Raster Tools > Correlate > Match
Command: IMATCH

You can use the **Match** tool to correlate a raster image by matching points between two images or between an image and the vector (CAD) data in the drawing. The **Match** tool moves, rotates, and scales the selected raster images so that the points specified on the source image coincide with the points specified on the destination data.

To correlate a raster image using the **Match** tool, choose it from the **Correlate** panel of the **Raster Tools** tab; you will be prompted to specify the source point1 on the raster to be correlated. Click on the required point to specify the first source point; you will be prompted to specify the destination point corresponding to the first point. Click on the drawing to specify the destination point; you will be prompted to specify the second source point. Select the second source point by clicking on the required point on the raster to be correlated; you will be prompted to specify the corresponding destination point. Click on the drawing to specify the destination point; the selected raster will be moved, scaled, and rotated so that the specified source and destination points match each other. Figure 2-24 shows the process of correlating a raster image using the **Match** tool.

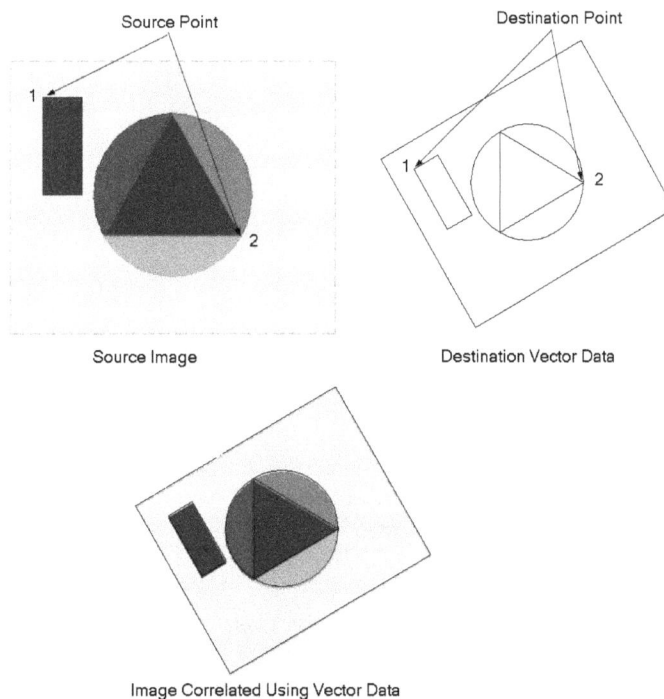

Figure 2-24 *Correlating the raster image using the* **Match** *tool*

Rubbersheeting an Image

Ribbon: Raster Tools > Correlate > Rubber Sheet
Command: IRSHEET

Sometimes you need to correct your raster images to remove distortion from the image. Distortion in the image can be parallax error, lens distortion, and unevenness of the terrain. Rubbersheet uses control points to match the source point and destination point in the drawing to remove such distortion.

The image correlation tools such as the **Displace**, **Match**, and **Scale** are useful in correlating images but are unable to remove distortion in the image. You can use the **Rubber Sheet** tool in AutoCAD Raster Design to correlate the inserted raster image in the drawing. Using this tool for image correlation, you can rectify most of the distortions in an image.

To correlate a raster image, choose the **Rubber Sheet** tool from the **Correlate** panel of the **Raster Tools** tab; the **Rubbersheet - Set Control Points** dialog box will be displayed, as shown in Figure 2-25.

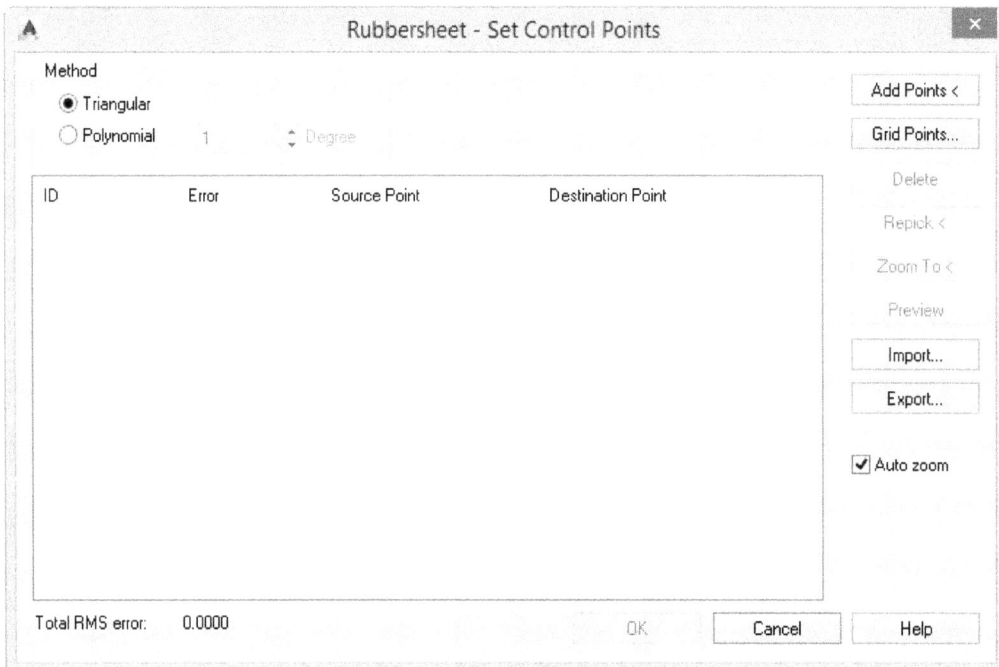

Figure 2-25 The **Rubbersheet - Set Control Points** *dialog box*

Using the options in this dialog box, you can specify the set of points that provide the matching points between the image and the drawing. These matching points, known as control points, can be selected by picking source and destination points in the drawing. They can also be specified by using a grid of destination points. After specifying the control points, you will have to select the method of image transformation. AutoCAD Raster Design offers two methods for image transformation: **Triangular** method and **Polynomial** method. On selecting the transformation method, AutoCAD Raster Design will use the specified control points and selected image transformation method to rubbersheet the image.

The options in the **Rubbersheet - Set Control Points** dialog box are discussed next.

Method Area

The **Method** area in the **Rubbersheet - Set Control Points** dialog box contains options for selecting the image transformation method. As explained earlier, AutoCAD Raster Design can transform an image using the Triangular or Polynomial transformation method. To choose the Triangular image transformation method, select the **Triangular** radio button in this area or choose the **Polynomial** radio button to use the Polynomial image transformation method. These transformation methods are briefly discussed next.

Triangular Image Transformation

In the Triangular image transformation method, AutoCAD Raster Design creates triangular regions within the raster image using the control points specified by the user. To create the triangular regions, AutoCAD Raster Design uses the Denaulay Triangulation method. Next, transformation is applied to each triangular area resulting in a more accurate transformation than the polynomial image transformation method.

This method applies transformation to only that part of the image which lies within the area defined by the Denaulays triangulation. The portion of the image lying outside the triangulation is not transformed. As a result, AutoCAD Raster Design discards the image data outside the triangulated area which leads to the loss of data. To minimise the loss of data, ensure that the control points are placed as close to the image extents as possible.

Polynomial Image Transformation

The Polynomial image transformation method transforms the entire image by matching the source and destination points as close as possible. As the transformation is applied on the entire image, this method does not result in loss of image data. However, this method does not always result in perfectly matched source and destination points. The positional error between the actual destination point (destination point on transformed image) and the destination point specified by the user is given by:

$$\text{Positional error} = \sqrt{(\Delta x + \Delta y)}$$

You can specify the polynomial degree in the **Degree** edit box corresponding to the **Polynomial** radio button. You have to specify more numbers of control points in the entire image if higher polynomial order is selected. The minimum number of control points for polynomial orders are as follows:

Polynomial Orders	Minimum Number of Control Points
1	3
2	6
3	10
4	15
5	21

Note

*The **Degree** edit box will be activated only when you select the **Polynomial** radio button and had specified minimum six control points.*

Note that, a higher polynomial degree will increase the positional accuracy of the image points at the matching locations (source point and destination points).

Add Points

You can choose the **Add Points** button to specify the control points by selecting the source and destination points. On choosing this button, the **Rubbersheet - Set Control Points** dialog box will be closed and you will be prompted to select the first source point. Select the source point on the image by clicking at the required point; you will be prompted to specify the destination point for the selected source point. Click on the required location in the drawing; you will be prompted to select the next source point. Continue specifying the required number of control points. To end specifying control points, press ENTER; the **Rubbersheet - Set Control Points** dialog box will displayed.

Note that the control point table in the dialog box displays ID of the control points, coordinates of the source and destination points in the control point, and positional error of the control points. Figure 2-26 displays a list of control points in the **Rubbersheet - Set Control Points** dialog box.

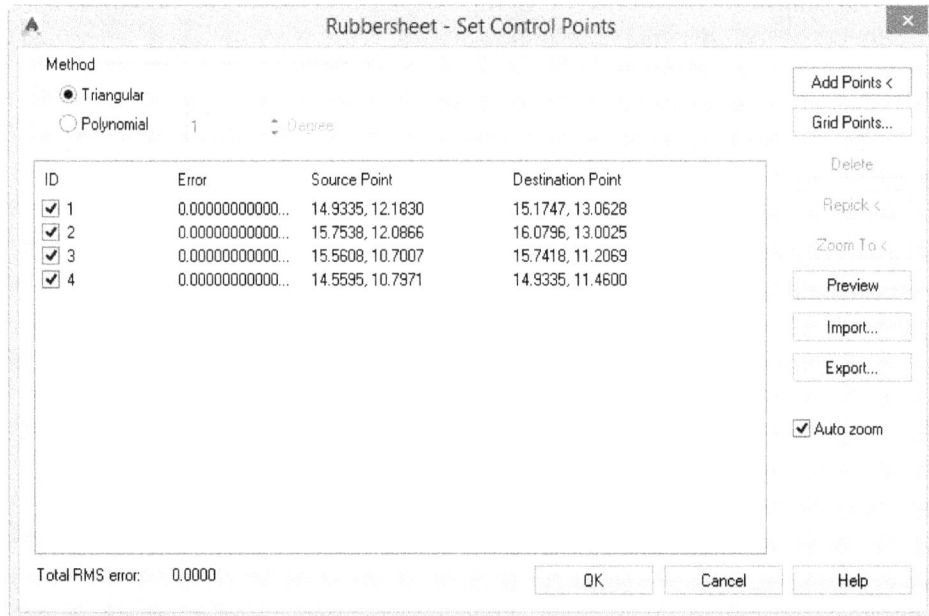

*Figure 2-26 The **Rubbersheet - Set Control Points** dialog box displaying list of control points*

Grid Points

You can choose the **Grid Points** button from the **Rubbersheet - Set Control Points** dialog box to invoke the **Grid Parameters** dialog box, as shown in Figure 2-27. Using the options in this dialog box, you can create a grid of destination points and then use them to specify the control points for image transformation.

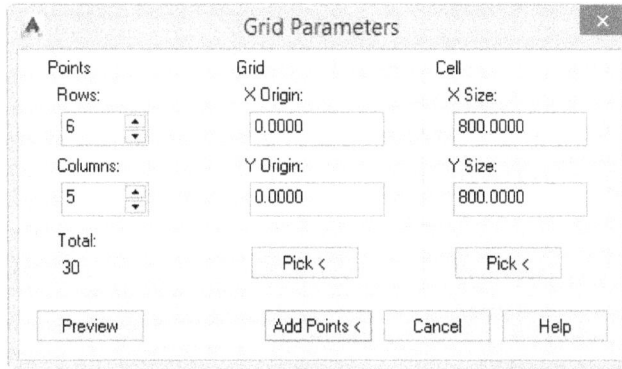

*Figure 2-27 The **Grid Parameters** dialog box*

Specifying the Control Points Using the Grid of Destination Points

In this dialog box, specify the required number of rows and columns in the **Rows** and **Columns** edit boxes, respectively. Note that on doing so, the **Total** option below the **Column** edit will display the total number of control points.

Next, specify the x and y coordinates of the grid origin in the **X Origin** and **Y Origin** edit boxes, respectively. Alternatively, you can specify the grid origin by picking a point in the drawing. To do so, choose the **Pick** button in the **Grid** area of the dialog box. On choosing this button, the **Grid Parameters** dialog box will close and you will be prompted to specify the grid origin. Click on the required point in the drawing; the **Grid Parameters** dialog box will be displayed again. Note that the **X Origin** and **Y Origin** edit boxes in the dialog box now display the coordinates of the selected point.

Next, specify the values for the column and row spacing of the grid in the **X Size** and **Y Size** edit boxes, respectively. Alternatively, you can graphically specify the row and column spacing in the grid. To do so, choose the **Pick** button in the **Cell** area of the dialog box; the **Grid Parameters** dialog box will close and you will be prompted to specify the first corner point of the grid cell. Click on the drawing to specify the first corner point. On doing so, you will be prompted to specify the second corner point. Click on the drawing to specify the point. On doing so, the **Grid Parameters** dialog box will be displayed again. Note that the **X Size** and **Y Size** edit boxes in the dialog box will now display the values corresponding to the dimensions of the cell size specified graphically.

After specifying the parameters for creating the grid, you can preview the grid by choosing the **Preview** button. Next, to specify the control points using the points in the grid as the destination points, choose the **Add points** button in the **Grid Parameters** dialog box; the dialog box will close and you will be prompted to specify the source point corresponding to the selected destination point in the grid. Select the source point on the image by clicking on the required point; AutoCAD Raster design will select the next point in the grid and will prompt you to select the corresponding source point.

Continue specifying the source point for all the destination points in the grid. After the source points for the destination points in the grid are specified, the **Rubbersheet - Set Control Points** dialog box will be displayed again.

Delete
You can choose the **Delete** button to remove (delete) a control point from the **Rubbersheet - Set Control Points** dialog box. Note that the **Delete** button will be activated only after you select a control point from the control point table.

Repick
You can choose the **Repick** button to edit an existing control point. To do so, select the required control point from the control point table in the **Rubbersheet - Set Control Points** dialog box; the **Repick** button will be activated. Choose this button; the **Rubbersheet - Set Control Points** dialog box will be closed and you will be prompted to specify the source point for the selected control point. Click on the drawing to specify the source point for the control point; you will be prompted to specify the destination point for the control point. Click on the drawing to specify the destination point; the **Rubbersheet - Set Control Points** dialog box will be displayed with the updated source and destination point for the selected control point.

> **Tip**
> *To edit multiple control points, select the required control points in the **Rubbersheet - Set Control Points** dialog box and then choose the **Repick** button. On doing so, you will be prompted to specify the source and destination points for the selected control points.*

Preview
You can preview the result of rubbersheet by choosing the **Preview** button from the **Rubbersheet - Set Control Points** dialog box. On choosing the button, the dialog box will close and the preview of the rubbersheet will be displayed in the drawing window. To end the display of preview, press ENTER.

Import
You can import a set of control points that are saved in a text (*.txt*) file. To do so, choose the **Import** button from the **Rubbersheet - Set Control Points** dialog box; the **Import** dialog box will be displayed. Use the options in this dialog box to browse and choose the text file that contains list of the saved control points. Next, choose the **Open** button; the **Import** dialog box will close and list of control points will be displayed in the control points table of the **Rubbersheet - Set Control Points** dialog box.

Export
You can also save the control points that are specified in the control points table of the **Rubbersheet - Set Control Points** dialog box as a text file. To do so, choose the **Export** button;

the **Export** dialog box will be displayed. In this box, specify the path and name of the text file into which you want to export the control points and then choose the **Export** button, the control points will be exported to the specified file.

Auto zoom

You can select the **Auto zoom** check box to enable AutoCAD Raster design to zoom to the selected control point on choosing the **Repick** button.

After specifying the required control points, choose the **OK** button; the **Rubbersheet - Set Control Points** dialog box will be closed and the image will be transformed using the specified options.

TUTORIAL

Before starting the tutorial you need to download and save the tutorial files on your computer. To do so, follow the steps given below:

1. Navigate to C drive in your system and create a folder with the name *AutoCADRasterDesign2016*.

2. Download the *c02_rd_2016_tut.zip* file from *http://www.cadcim.com*. The path of the file is as follows:
    Textbooks > Civil/GIS > AutoCAD Raster Design > Exploring AutoCAD Raster Design 2016

3. Now, save and extract the downloaded folder at the following location:

    C:\AutoCADRasterDesign2016

Notice that *c02_rd_2016_tut* folder is created within the *AutoCADRasterDesign2016* folder.

Tutorial 1	Insert and Rubbersheet Raster Image

In this tutorial, you will start the AutoCAD Raster Design application and then insert a raster image using the **Insert** tool. Figure 2-28 shows the raster image to be inserted. Next, you will specify the control points for rubbersheeting the raster image. You will also export list of the specified control points as a text file. **(Expected time: 30 min)**

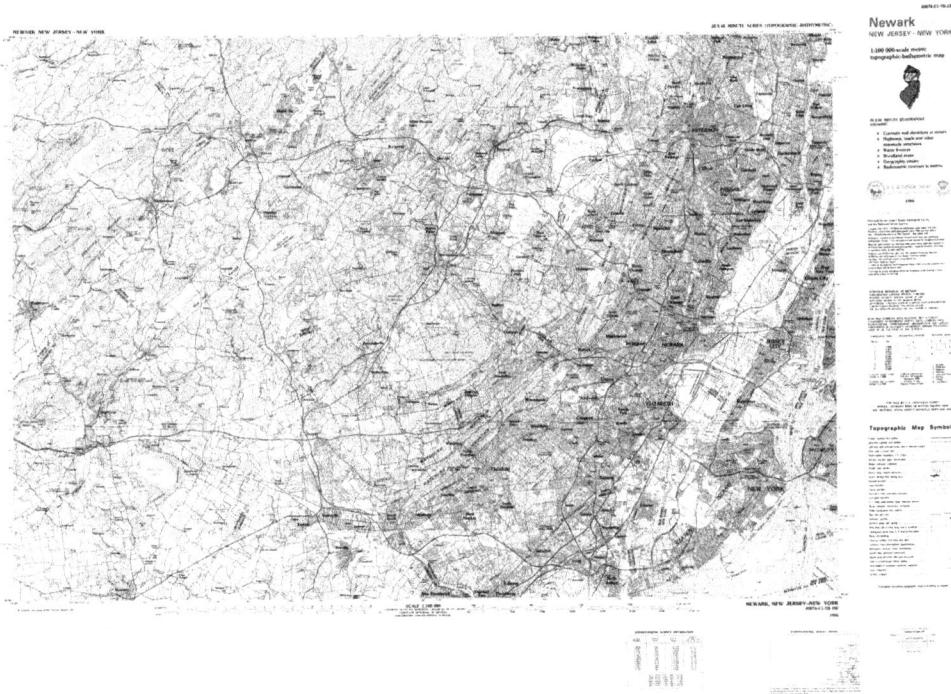

Figure 2-28 *The Newark (New Jersey-New York) area toposheet*

The following steps are required to complete this tutorial:

a. Start the base application and open the drawing file.
b. Insert the raster image.
c. Rubbersheet the raster image.
d. Save the drawing file and the correlated raster image.

Starting the Base Application and Opening the Drawing File

1. Start AutoCAD Raster Design and then choose the **Open** button from the Application menu; the **Select File** dialog box is displayed.

2. In this dialog box, browse to the location *C:\AutoCADRasterDesign2016\c02_rd_2016_tut\ c02_tut01*, select the **c02_Tut01.dwg** file and then choose the **Open** button. Drawing file with **UTM27-18** coordinate system assigned is opened.

Inserting the Raster Image

In this section you will learn to insert a raster image into the drawing using the **Insert** tool of AutoCAD Raster Design.

1. Choose the **Insert** tool from the **Insert & Write** panel in the **Raster Tools** tab; the **Insert Image** dialog box is displayed.

2. In this dialog box, select the **Insertion dialog** radio button from the **Insert Options** area.

3. Next, browse to the location *C:\AutoCADRasterDesign2016\c02_rd_2016_tut\c02_tut01* from **Look in** text box and then choose the **NewJersey.jpg** file.

4. Next, choose the **Open** button; the **Insert Image** dialog box is closed and the **Image Insertion** dialog box is displayed.

5. Ensure that the **Insertion** tab is chosen.

6. In this dialog box, specify the correlation parameters, as shown in Figure 2-29.

*Figure 2-29 The **Image Insertion** dialog box with the parameters specified for image insertion*

7. Choose the **OK** button; the **Image Insertion** dialog box is closed and the raster image is inserted in the drawing using the specified settings.

8. Next, enter **Z** in the Command prompt area and press ENTER; you are prompted to specify the zoom option.

9. Specify **E** and then press ENTER; the drawing will zoom to its extents.

Note that the raster image is a topomap of the New Jersey area. It also has various markings showing the map grid. You will use the intersection of the 10000 meter grid lines (labeled in red colored numbers) to specify the control points for transforming the raster image.

Note

*You can also use the **Extents** tool from the Navigation Bar to zoom to the extent of* ⊠ Extents *the drawing.*

Rubbersheeting the Raster Image

In this part of the tutorial, you will rubbersheet the raster image using the Polynomial image transformation method. Note that you will use the grid to specify the control points for transforming the image.

1. Choose the **Rubber Sheet** tool from the **Correlate** panel of the **Raster Tools** tab; the **Rubbersheet - Set Control Points** dialog box is displayed.

2. Choose the **Grid Points** button; the **Rubbersheet - Set Control Points** dialog box is closed and the **Grid Parameters** dialog box is displayed.

3. In this dialog box, enter value for the different parameters in their corresponding edit boxes, as shown in Figure 2-30.

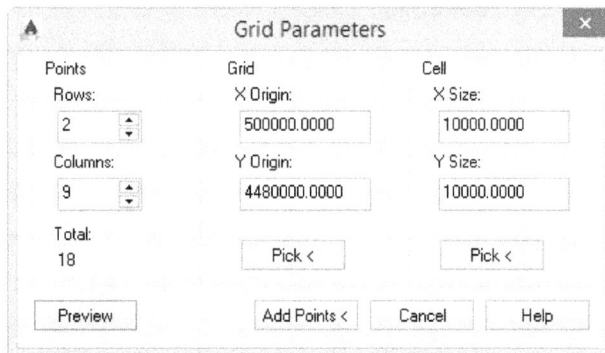

*Figure 2-30 The **Grid Parameters** dialog box with the parameters for creating grid*

4. Next, choose the **Preview** button in the **Grid Parameters** dialog box; the dialog box is closed and preview of the grid is displayed in the drawing area as shown in Figure 2-31.

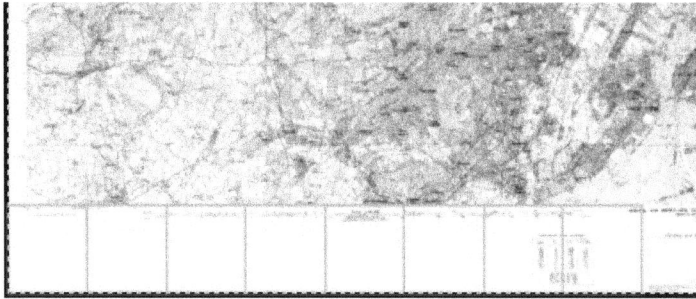

Figure 2-31 *The preview of the grid displayed in the drawing area*

Note

*The grid is displayed in white color by default. You can change the grid color by choosing **Color of a layer** from the **AutoCAD Layers** panel of the **Home** tab.*

5. Press the ESC key to close the preview mode of the grid.

    Now you will specify the control points for rubbersheeting the raster image.

6. Choose the **Add Points** button; the **Grid Parameters** dialog box is closed and you are prompted to specify the source point as 0,0. Notice that the destination point is at the bottom left corner of the grid and the cursor is attached to this point, as shown in Figure 2-32.

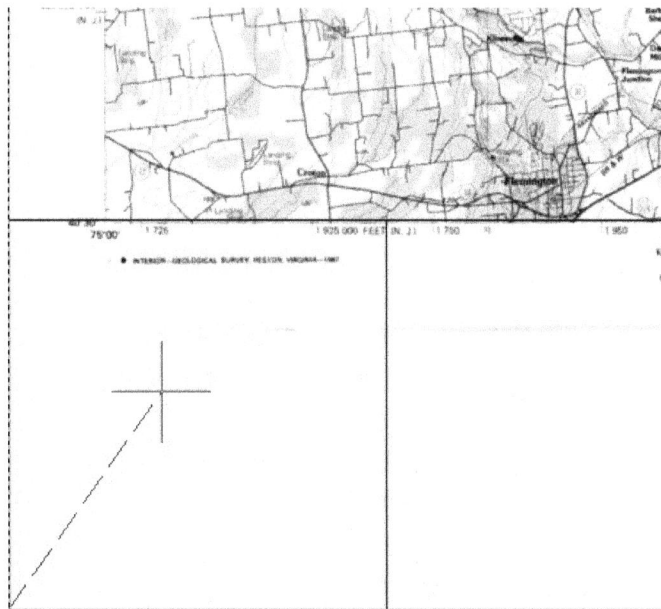

Figure 2-32 *The cursor attached to the destination point on the grid*

7. Move the cursor to the grid intersection 1 indicated by red color on the raster image. Figure 2-33 shows an enlarged view of the raster at grid intersection 1. Next, click on the intersection; the cursor now snaps to the next point on the grid and you are prompted to specify the source point for the selected point.

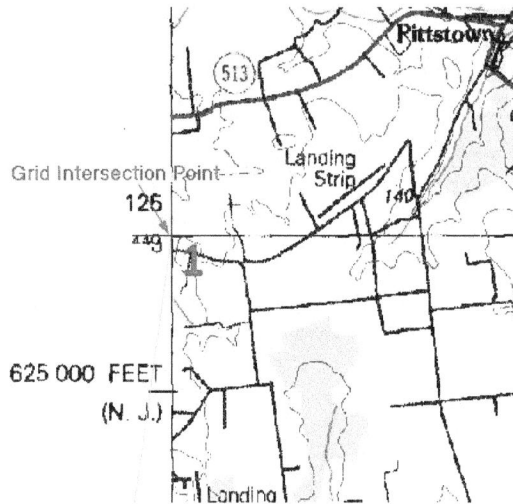

Figure 2-33   *The enlarged view of the raster image at grid intersection 1*

Tip
For precisely specifying the source point on the raster image, zoom in to the drawing. You can use the mouse roller to zoom in and out of the drawing.

Note
The source points should be specified with utmost care and precision as they affect the accuracy of the result while rubbersheeting.

8. Similarly, click on the grid intersection 2 on the raster, as shown in Figure 2-34; you will be prompted to specify the source point for the next grid point.

Figure 2-34   *The enlarged view of the raster image at grid intersection 2*

9.   Continue specifying source points from grid intersection marked 3 to grid intersection marked 18. On specifying the 18th intersection point, the **Rubbersheet- Set Control Points** dialog box is displayed with specified control points, as shown in Figure 2-35.

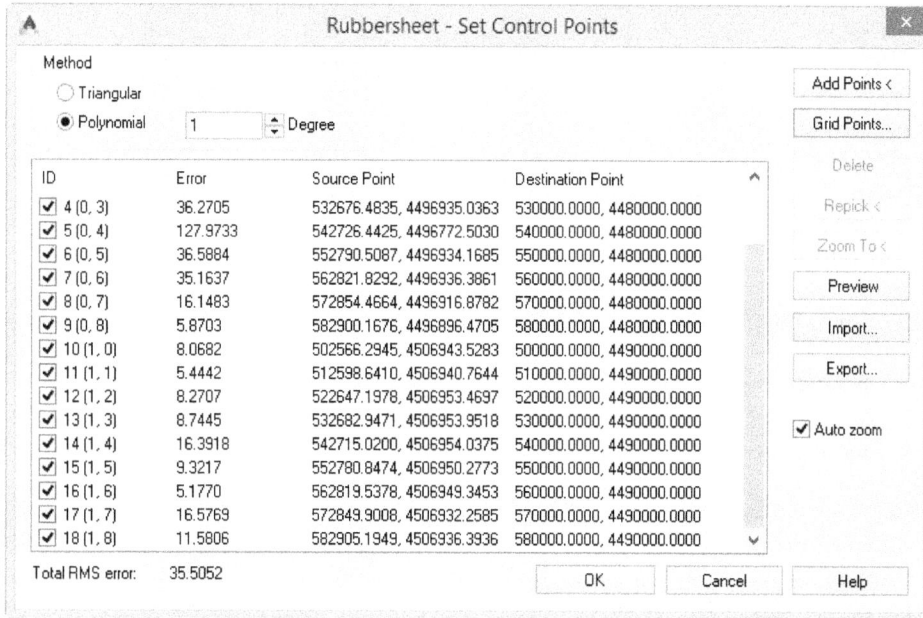

*Figure 2-35 The **Rubbersheet - Set Control Points** dialog box with the list of specified control points*

Note that the dialog box now displays the list of the control points and their parameters.

Note
*The **Total RMS error** depends on the users precision on specifying the points in the image.*

Next, you will export the specified control points as an text file.

10. Choose the **Export** button; the **Export** dialog box is displayed.

11. In this dialog box, browse to the location *C:\AutoCADRasterDesign2016\c02_rd_2016_tut\ c02_tut01*.

12.   Next, specify the name **NewJerseyControlPoints** in the **File name** edit box and then choose the **Export** button; the **Export** dialog box is closed and the control points are exported as a text file to the specified location and the **Rubbersheet - Set Control Points** dialog box is displayed.

13. In the **Rubbersheet - Set Control Points** dialog box, make sure that the **Polynomial** radio button is selected and then choose the **OK** button; the dialog box is closed and toposheet is rubbersheeted.

Note

The process of rubbersheeting is a resource intensive process and may take some time depending on the configuration of your computer. The progress of the process is displayed at the bottom right of the user interface.

Saving the File

1. Choose the **Save As** option from the **Application Menu**; the **Save Drawing As** dialog box is displayed.

2. In this dialog box, browse to the following location:

   C:\AutoCADRasterDesign2016\c02_rd_2016_tut\c02_tut01

3. In the **File name** edit box, enter **c02_Tut01_Result**.

4. Choose the **Save** button; the dialog box is closed and drawing file is saved with the name *c02_Tut01_Result.dwg* at the specified location. Also, the **Save Image** dialog box is displayed, as shown in Figure 2-36, informing you that the raster image in the drawing has been modified.

*Figure 2-36 The **Save Image** dialog box*

5. Choose the **Save As** button from the dialog box; the **Save As** dialog is displayed.

6. In this dialog box, browse to the following location:

   C:\AutoCADRasterDesign2016\c02_rd_2016_tut\c02_tut01

7. In the **File name** edit box, enter **NewJerseyCorrelated**.

8. Ensure that the **JPEG File Interchange Format (*.jpg, *.jpeg)** option is selected from the **Files of types** drop-down list and then choose the **Save** button; the **Save As** dialog box is closed and the **Encoding Method** dialog box is displayed, as shown in Figure 2-37.

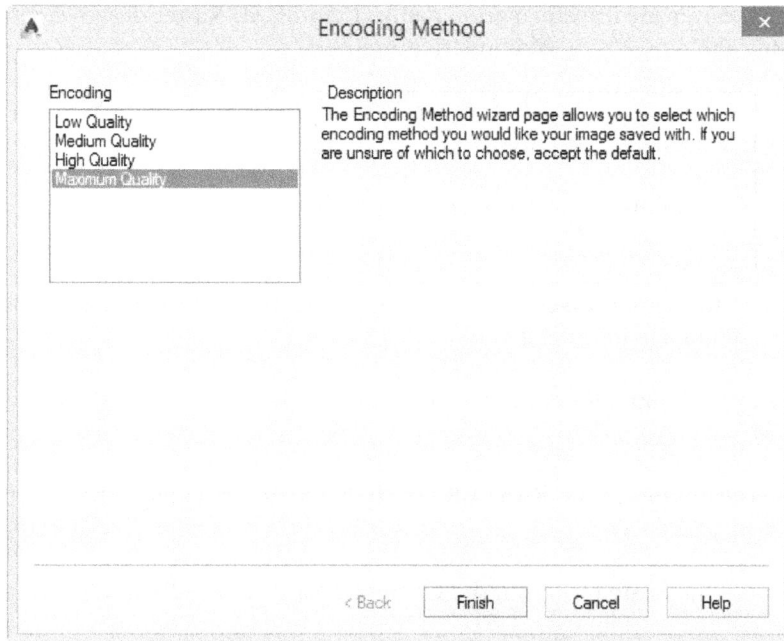

*Figure 2-37 The **Encoding Method** dialog box*

9. Choose the **Maximum Quality** option from the **Encoding** list in this dialog box and then choose the **Finish** button; the raster image is exported to the specified location.

Self-Evaluation Test

Answer the following questions and then compare them to those given at the end of this chapter:

1. Which of the following options can be chosen to insert a raster image into the drawing?

    (a) **Quick insert**   (b) **Insertion wizard**
    (c) **Insertion dialog**   (d) All of the above

2. Which of the following correlation methods result in the loss of image data due to image transformation?

    (a) **Triangular**   (b) **Match**
    (c) **Polynomial**   (d) All of the above

3. The second degree Polynomial image transformation method requires a minimum of _____ number of control points to transform a raster image.

4. In the Triangular image transformation method, AutoCAD Raster design creates triangular regions using the _____ triangulation method.

5. You can choose the _____ button from the **Rubbersheet - Set Control Points** dialog box to edit an existing control point.

6. You can use the options in the **Grid Parameters** dialog box to specify the parameters for creating grid. (T/F)

7. You can select the **Quick insert** radio button in the **Insert Image** dialog box to directly insert a selected image into the given drawing. (T/F)

8. AutoCAD Raster design allows you to adjust the brightness, fade, and contrast of a raster image. (T/F)

9. You can use the options in the **RASTER DESIGN** palette to manage the inserted rasters in your drawings. (T/F)

10. You can choose the **Scale** tool from the **Correlate** panel of the **Raster Tools** tab to change the scale of the inserted image. (T/F)

Review Questions

Answer the following questions:

1. The **RASTER DESIGN** palette display a list of images in the drawing in a _____ view.

2. To edit multiple control points in the **Rubbersheet - Set Control Points** dialog box you need to choose the _____ button.

3. You can save the control points as a text file by choosing the _____ button from the **Rubbersheet - Set Control Points** dialog box.

4. The **Item view** in the **RASTER DESIGN** palette displays the properties of the selected image in a table. (T/F)

5. You can use the **Pick** button from the **Grid Parameters** dialog box to specify the control points using the points in the grid as the destination points. (T/F)

6. AutoCAD Raster Design does not facilitate you to import control points that are defined in an external text (*.txt*) file. (T/F)

7. You need to select an image to scale by clicking on its image frame. (T/F)

8. You can select a point on the image frame by using the **Object Snap** option. (T/F)

EXERCISE

Exercise 1

Download the *c02_rd_2016_exr.zip* from *http://www.cadcim.com* and rubbersheet the *exr_2.jpg* raster image, refer to Figure 2-38.   **(Expected time: 30 min)**

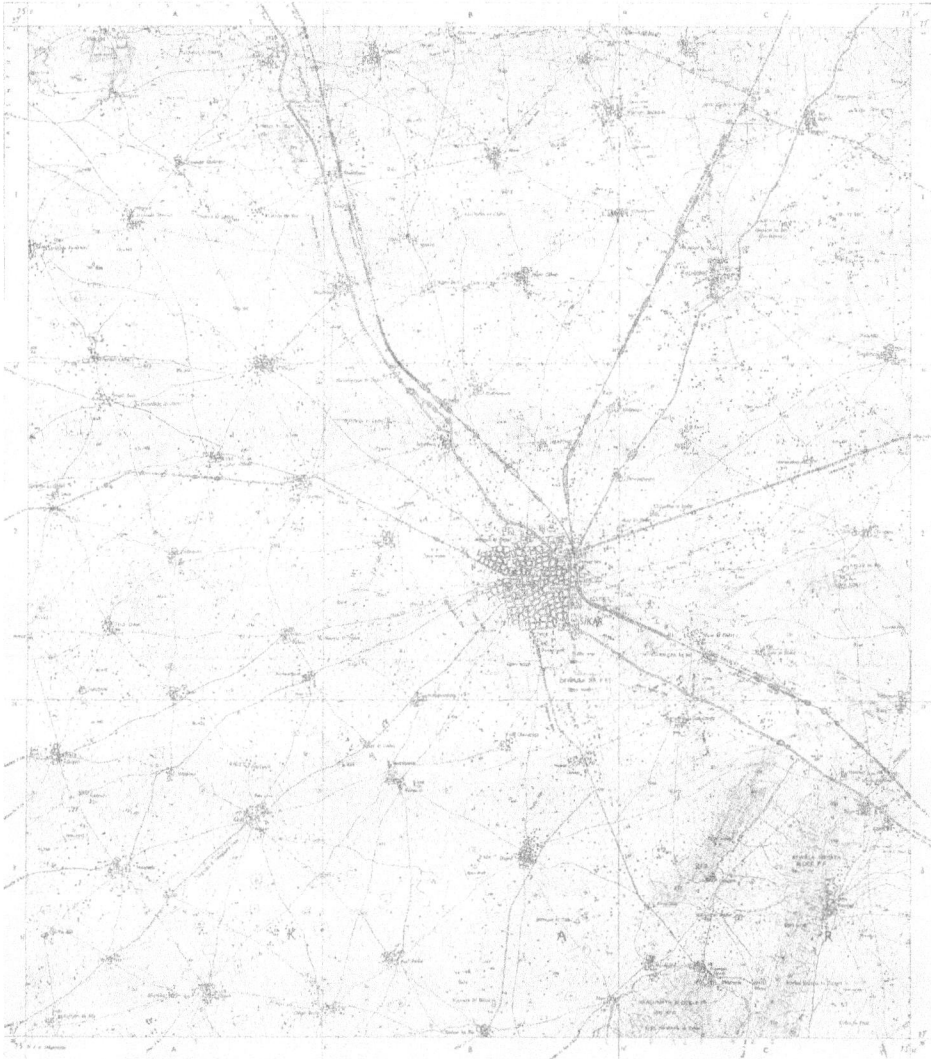

Figure 2-38 *Image on which rubbersheet has to be performed*

Answers to Self-Evaluation Test

1. d, **2.** a, **3.** six, **4.** Denaulay, **5. Repick <**, **6.** T, **7.** T, **8.** T, **9.** T, **10.** T

Chapter 3

Image Management Tools

Learning Objectives

After completing this chapter, you will be able to:

- *Clean up an image*
- *Crop an image*
- *Touch up an image*
- *Raster snap an image*

INTRODUCTION

AutoCAD Raster Design is designed to process raster data and allows the user to modify digital images for better visualization and understanding. In this chapter, you will learn to clean up an image by removing bias, speckle, and skew from the image. You will also learn to crop raster data within an image or in multiple images.

In addition, this chapter discusses about the image touchup tools, remove tools, and raster snap tools. These tools help the analyst to provide more accurate raster data for image interpretation. These tools and their functions are discussed briefly in the forthcoming sections.

CLEANING UP AN IMAGE

AutoCAD Raster Design provides features to clean up raster data. You can use various image clean up tools to remove errors from an image. Sometimes, raster images are damaged and produce error while scanning or printing. You need to remove these errors to get error free raster data. Using powerful drawing clean up tools, you can remove these errors from the images. Different types of clean up tools are shown in Figure 3-1. These cleanup tools are discussed next.

Figure 3-1 *Various clean up tools*

Deskew Tool

Ribbon: Raster Tools > Edit Panel > Cleanup drop-down > Deskew
Command: IDESKEW

The **Deskew** tool is used to rotate an image. To deskew a raster image, choose the **Deskew** tool from the **Cleanup** drop-down in the **Edit** panel of the **Raster Tools** tab; you will be prompted to select base point of the image. Next, select the base point of the image to rotate. Now you can define the source angle by entering a value or by picking two points one by one on the raster image. Also you need to define the destination angle by typing a value in the command line or by picking two points on the image. Different types of geometric distortion in an image are shown in Figure 3-2.

Note
*While performing any operation using the tools from the **Raster Tools** tab if a drawing contains more than one raster image, you will be prompted to select the image. Select the image; you will be prompted to specify the base point of the selected image.*

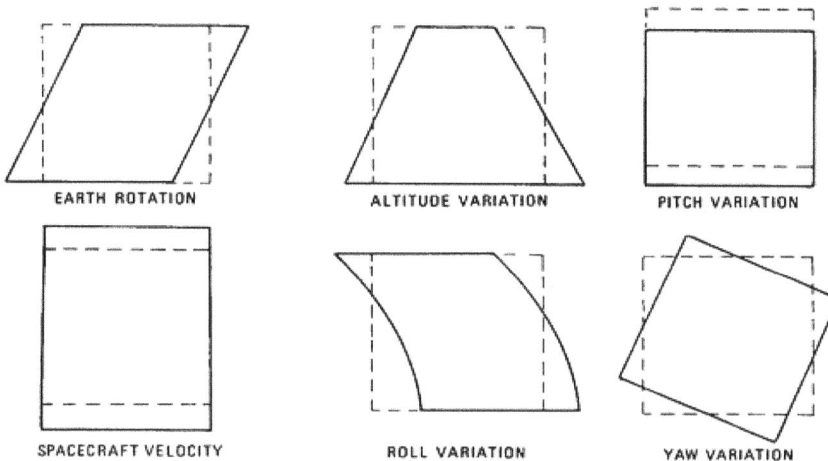

Figure 3-2 *Geometric distortions in an image*

Despeckle Tool

Ribbon:	Raster Tools > Cleanup drop-down > Despeckle
Command:	IDESPECKLE

Sometimes in images you find small marks or stray dots also termed as speckles. These marks or dots are considered as unwanted parts in an image and therefore need to be removed. You can use the **Despeckle** tool to remove these errors from the entire image or from a small portion by means of a polygon, clip region, or an existing vector entity.

To remove speckles from an image, choose the **Despeckle** tool from the **Cleanup** drop-down. Select the image from which you need to remove speckles. You can remove speckle from the entire image or from any region. Next, to specify the speckle size from the image by selecting the speckle or by defining the size in the Command prompt area. After selecting the speckle or specifying the size, the desired speckles will be highlighted in the drawing window. If you want to delete the highlighted speckles from the drawing window, press ENTER.

Note
While determining the size to remove a speckle, you should be careful that a large size can remove parts of the text from the image.

Bias Tool

Ribbon:	Raster Tools > Cleanup drop-down> Bias
Command:	IBIAS

An image may be distorted due to paper shrinkage or due to optical distortions in the scanning process. This error can be resolved by resizing the image in one or two dimensions. The **Bias** tool corrects distortions in the aspect ratio of the image.

To remove this bias, choose the **Bias** tool from the **Cleanup** drop-down; you will be prompted to select an image. Select the image and then click on a point in the drawing window to specify the base point of the image. After specifying the base point in the drawing, specify the calibration distance for the X direction by clicking on two points or by entering the desired values in the Command prompt area. Similarly, specify the distance for the Y direction; the image will be resized. Figure 3-3 shows the corrected biased image.

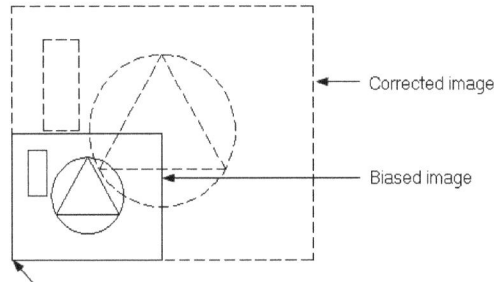

Figure 3-3 Corrected biased image

Invert Tool

Ribbon:	Raster Tools > Cleanup drop-down > Invert
Command:	IINVERT

In AutoCAD Raster Design, you can invert light and dark shades in an image. You can also invert bitonal, color, and grayscale images using the **Invert** tool. To invert the shades from an image(s), choose the **Invert** tool from the **Cleanup** drop-down. Next, select the image or images to start the inversion process. After the inversion process is finished, the color of the image or images will be inverted.

Tip
You can change the background color of the drawing window to extract better information from the image. You can also set the transparency of the background according to your need.

Mirror Tool

Ribbon: Raster Tools > Cleanup drop-down > Mirror
Command: IMIRROR

While scanning an image, scanning software might ocassionally scan the back of the image and change the original orientation. You can correct such errors by using the **Mirror** tool. After scanning operation, the image orientation is generally mirrored. To remove such errors, choose the **Mirror** tool from the **Cleanup** drop-down; the **Mirror** dialog box will be displayed with the **Top To bottom** and **Side To side** radio buttons; refer to Figure 3-4. If you want to flip the image along the horizontal axis, choose the **Top To bottom** radio button and if you want to flip the image along the vertical axis, choose the **Side To side** radio button.

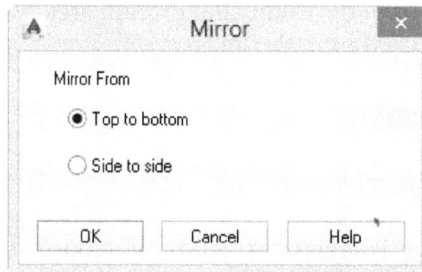

Figure 3-4 *The* *Mirror* *dialog box*

Note
If the drawing contains more than one raster image, you will be prompted to select the image to be mirrored.

CROPPING AN IMAGE

AutoCAD Raster Design has various tools to remove any unwanted region from your drawing. Cropping an image preserves the selected image area and removes the unwanted image area from the drawing and reduces the file size.

You can crop a raster image or multiple images using various cropping tools located in the **Crop** drop-down of the **Raster Tools** tab, refer to Figure 3-5. These tools are discussed next.

Figure 3-5 *The* *Crop* *tools drop-down*

Line Tool

Ribbon: Raster Tools > Crop drop-down > Line
Command: ICROPLINE

The **Line** tool crops entire image outside the specified linear region. To crop the desired area, choose the **Line** tool from the **Crop** drop-down; you will be prompted to specify the

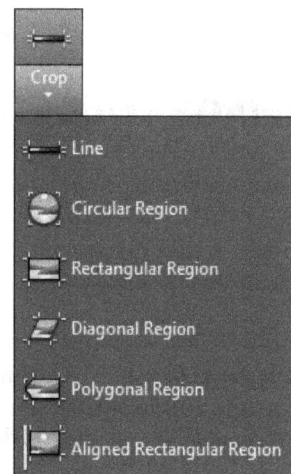

first point of the line. Specify the first point; you will be prompted to specify the second point of the line. After specifying the second point of the line, the entire image will be cropped retaining the image as per the specified width of the line segment.

Circular Region Tool

Ribbon: Raster Tools > Crop drop-down > Circular Region
Command: ICROPCIRC

The **Circular Region** tool crops the image outside the defined circular region. To crop the region outside the defined circle, choose the **Circular Region** tool from the **Crop** drop-down. Next, create a circular region using the options in the Command prompt area; the image will be cropped and the unwanted area will be removed from the image.

Polygonal Region Tool

Ribbon: Raster Tools > Crop drop-down > Polygonal Region
Command: ICROPPOLY

The **Polygonal Region** tool crops the image outside the defined polygonal region. To crop a desired polygonal region from the image, choose the **Polygonal Region** tool from the **Crop** drop-down; you will be prompted to specify the polygonal region by specifying at least three points on the desired raster image area. Next, specify the polygonal region and then press ENTER to close the polygon; the region will be cropped and the unwanted area of the image will be removed from the drawing window.

> **Note**
> *You can save the cropped image by choosing the **Save** button from the **Insert** & **Write** panel of the **Raster Tools** tab.*

IMAGE TOUCHUP

In AutoCAD Raster Design, you can correct imperfections in scanned bitonal images or drawings. Image touchup tools can be used to improve usability and legibility of scanned images. With the help of these tools, you can reduce the time and effort in cleaning up scanned images and maps. Using multiple resizable brushes, you can fill the background and foreground colors of the drawing image.

AutoCAD Raster Design offers various touchup tools such as **Square brush**, **Round brush**, and **Diagonal brush** in the **Touchup** toolbar. You can change the size of the touchup brush to match the line weight. Brush size also varies according to the zoom level. You can also interchange the color of the image and the background. Also you can change the direction of the lines or polylines.

Touchup Tools

Ribbon: Raster Tools > Touchup
Command: ITOUCHUP

The touchup tools are used to reform inaccurate lines or polylines on the drawing image. To invoke this tool, zoom in on the image so that the image pixels are clearly visible. Next, choose the **Touchup** tool from the **Edit** panel of the **Raster Tools** tab; the **Touchup** toolbar will be displayed. Choose the desired tool such as **Square brush**, **Round brush**, or **Diagonal brush** from the **Touchup** toolbar. Then, click and drag the cursor in the desired direction; the object will be modified. You can also modify the brush size by using the **Resize brush** tool. If you are satisfied with the modified line or polyline after editing, choose the **OK** button in the **Touchup** toolbar to end the session. The **Touchup** toolbar and its tools are shown in Figure 3-6.

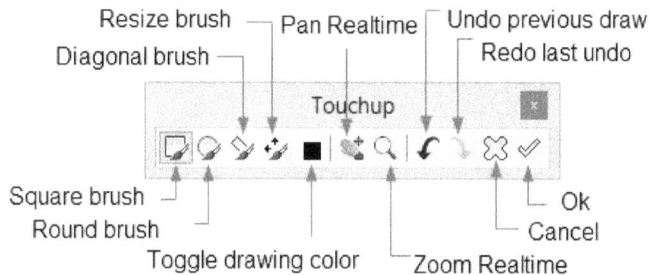

Figure 3-6 The **Touchup** toolbar

REMOVING (RUBBING) IMAGES

Sometimes, you need to remove stains, tear marks, wrinkles, and tape shadows from the scanned images for better visualization of features. In AutoCAD Raster Design, you can remove such unwanted elements from an image by using various rubbing tools such as **Line**, **Circle**, **Arc**, and so on. These tools permanently change the image. These tools use the transparent color for rubbing process and give output in background color. If the background color is black, the rubbed area will be displayed in black color only and if you change the background color, the color of the rubbed area will also change according to the changed background color. These rubbing tools are discussed next.

> **Note**
> *You can use the **UNDO** command at any stage during the rubbing process for rubbing the changes done during the process.*

Tip
*Before using the remove tools, you should edit the value in the **Rub/crop width** edit box in the **Edit** drop-down of the **Raster Tools** tab.*

Line Tool

Ribbon: Raster Tools > Edit panel > Remove drop-down > Line
Command: IRUBLINE

The **Line** tool is used to remove features linearly from the raster image. This tool permanently removes the raster data information from the image. To remove a raster line from the image, choose the **Line** tool from the **Remove** drop-down of the **Raster Tools** tab. Next, specify the first point and the last point on the raster image; the line feature as per the default width mentioned in drafting settings will be removed from the image. You can save the image for further use by choosing the **Save** button in the **Insert & Write** panel of the **Raster Tools** tab.

Circle Tool

Ribbon: Raster Tools > Edit panel > Remove drop-down > Circle
Command: IRUBCIRCLE

The **Circle** tool is used to remove a circle(s) from the raster image. To remove a circle, choose the **Circle** tool from the **Remove** drop-down of the **Raster Tools** tab. Next, select the center point of the raster circle and then specify second point to define the radius of the raster circle. Next, press ENTER, the circle will be created on the existing circle in the drawing and the region within the specified crop width of the circle will be removed permanently from the image.

Arc Tool

Ribbon: Raster Tools > Edit panel > Remove drop-down > Arc
Command: IRUBARC

The **Arc** tool is used to remove unwanted arc(s) from the raster image. In this arc removal process, the image gets cropped as per the specified crop width of the arc. To remove arc entity, choose the **Arc** tool from the **Remove** drop-down in the **Raster Tools** tab. Select the start point on the existing arc entity in the raster image followed by the center point of the arc and then select the end point. The selected arc entity will be removed from the raster image.

Polyline Tool

Ribbon: Raster Tools > Edit panel > Remove drop-down > Polyline
Command: IRUBPLINE

In raster data, you may find some unwanted polylines. You can remove them by using the **Polyline** tool. To remove a polyline, choose the **Polyline** tool from the **Remove** drop-down in the **Raster Tools** tab. Select the first and last points of the polyline that you want to remove from the raster image. Next, right-click; a shortcut menu will be displayed, as shown in Figure 3-7. Choose the **Enter** option from the shortcut menu; the selected polyline objects will be removed from the image.

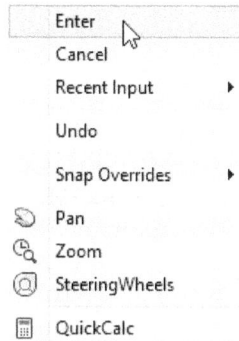

Figure 3-7 *The shortcut menu displayed to select the* **Enter** *option*

Circular Region Tool

Ribbon:	Raster Tools > Remove drop-down >Circular Region
Command:	IRUBCIRC

The **Circular Region** tool is used to remove a desired region from the raster image within a circular boundary. To remove an unwanted region, choose the **Circular Region** tool from the **Remove** drop-down and then select center point and radius for the circle on the desired area to be removed from the raster image. Next, right-click on the drawing window; a shortcut menu will be displayed. Choose the **Enter** option from the shortcut menu; the region within the circular boundary will be removed.

Note
Like the **Circular Region** *tool, the* **Rectangular Region** *and* **Diagonal Region** *tools are used to remove the desired region within the rectangular and diagonal boundaries, respectively.*

Polygonal Region Tool

Ribbon:	Raster Tools > Edit panel > Remove drop-down > Polygonal Region
Command:	IRUBPOLY

The **Polygonal Region** tool is used to remove a desired region from the raster image within a polygonal boundary. To do so, choose the **Polygonal Region** tool from the **Remove** drop-down and then select the three points on the desired area of the raster image. Next, right-click on the drawing window; a shortcut menu will be displayed. Choose the **Enter** option from the shortcut menu; the region within the polygonal boundary will be removed.

Note

*Once you choose the **Save** button to save an image, the image will be saved and you will not be able to retrieve the previous environment. Even the **UNDO** command will not work.*

Raster Under Vector Tool

Ribbon: Raster Tools > Edit panel > Remove drop-down > Raster Under Vector
Command: IRMVEXISTING

The **Raster Under Vector** tool is used to remove raster entities underlying within the vector geometry. This tool will only be effective if you draw a vector geometry on the raster image using AutoCAD tools. This tool removes the raster data outside the defined vector geometry depending on the specified rub/crop width. To remove a raster entity underlying a vector geometry, choose the **Raster Under Vector** option from the **Remove** drop-down. Next, select the vector geometry to remove the portion of the underlying raster image. On doing so, the underlying raster image outside the vector geometry will be removed.

Smart Pick Tool

Ribbon: Raster Tools > Edit panel > Remove drop-down > Smart Pick
Command: IDELSMART

The **Smart Pick** tool is used to remove raster lines, arcs, and circles from the bitonal image. To remove raster entities from the image, choose the **Smart Pick** tool from the **Remove** drop-down and then click on the entities to be removed; the selected entities will be removed.

Line Entity Tool

Ribbon: Raster Tools > Edit panel >Remove drop-down > Line Entity
Command: IDELLINE

The **Line Entity** tool is used to remove a particular raster line from the bitonal image(s). To remove a line feature from an image, choose the **Line Entity** tool from the **Remove** drop-down. Next, select a single point on the raster line or select the **2P** option from the Command prompt area; the line entity from the bitonal raster image will be removed.

Circle Entity Tool

Ribbon: Raster Tools > Edit panel > Remove drop-down > Circle Entity
Command: IDELCIRCLE

The **Circle Entity** tool is used to remove a particular circle entity from the image. To do so, choose the **Circle Entity** tool from the **Remove** drop-down in the **Raster Tools** tab. Next, select the **Centre**, **2P**, or **3P** option from the Command prompt area. Select the desired circle in the image to remove from the image; the circle entity will be removed from the bitonal image.

Arc Entity Tool

Ribbon: Raster Tools > Edit panel > Remove drop-down > Arc Entity
Command: IDELARC

The **Arc Entity** tool is used to automatically remove the intersection of the arc with other objects. To remove the arc object from the bitonal image, select the **Arc Entity** from the **Remove** drop-down. Next, select the center point or enter 3 points on the arc object to be removed; the object will be removed from the raster image.

RASTER SNAP OPTIONS IN RASTER DESIGN

In AutoCAD Raster Design, you can snap the cursor to a desired point in an object using the raster snap options. Raster snap options make it easy to select points on raster objects during the use of vectorization tools or REM commands. Raster snap options work only with bitonal images. The uses of raster snap options are similar to object snap modes (OSNAP) of AutoCAD, but they only snap to raster objects rather than vector entities. These options are used to move the cursor to the center, end, corner, intersection, or edge of a bitonal raster object. These options are located in the **Snap** drop-down of the **Raster Tools** tab. Note that, **Snap** drop-down will only be visible when the width of the Ribbon is greater than the width of the screen. When the width of the Ribbon is less as compared to width of the screen, the **Snap** drop-down changes into the **Snap** panel.

Raster Snap Menu

You can toggle these tools on or off by choosing the **Raster Snap** button from the **Snap** drop-down in the **Raster Tools** tab, refer to Figure 3-8. The raster snap tools are discussed next.

Figure 3-8 The **Raster Snap** menu

Center

This option is used to select the center of the raster line for raster entity manipulation or bitonal image editing.

End

This option is used to select the end of the raster line in an image for editing the raster image.

> **Note**
> *The **End** tool is used to select the intersection points of two line features.*

Corner

This option is used to snap the intersection of two raster lines.

Intersection

This option is used to snap the intersection of three or more raster lines.

Edge

This option is used to snap the edge of a raster line in the snap window.

Snap Override

In AutoCAD Raster Design, you can use raster snap override tools such as **Snap**, **Polar Tracking**, and **Ortho**. These options are available in the **Snap Override** drop-down in the **Snap** panel in the **Raster Tools** tab. The options are discussed next.

Snap

Choose the **Snap** option to activate the grid snap mode. If this mode is active, the cursor will move in fixed increments using the current settings of the snap grid. The F9 key acts as a toggle key to turn the snap mode on or off.

Polar Tracking

If you choose the **Polar Tracking** option, the movement of the cursor is restricted along a path determined by the angle set as the polar angle. You can also use the F10 key to toggle this option on or off.

Ortho

You can choose the **Ortho** option to activate or deactivate the ortho mode.

Show Aperture

You can select the **Show Aperture** button to display the aperture of the cursor while using raster snap.

Size

You can specify the aperture size in the **Size** edit box. The size varies from 5 to 300 pixels.

MERGE IMAGES AND VECTORS

In AutoCAD Raster Design, you can merge two or more raster images or vectors with the existing raster images. In this process, the source images or vectors get merged into a single image. When two or more satellite images are merged into a new or existing image then the process is called image merging. When two or more vector data are merged into a new or existing vector then the process is called as vector merging. Using different merge tools in AutoCAD Raster Design, you can combine number of rasters and vectors into a single frame. Merging of images does not affect the resolution. Merge tools are discussed next.

Merge Images Tool

Ribbon:	Raster Tools > Edit drop-down > Merge Images
Command:	IIMERGE

The **Merge Images** tool is used to merge different raster images into a new or existing image frame. To merge raster images into a single image frame, select the **Merge Images** tool from the **Edit** drop-down in the **Raster Tools** tab. On doing so, the cursor will change into a selection box and you will be prompted to select the source image(s). Select the images to be merged and press ENTER; you will be prompted to select the destination image. Select the destination image and press ENTER. Then, again you will be prompted to remove the source images after the merge. Select the **Yes** option from the Command prompt area; the merge process will start and the images will be merged.

Merge Vector Tool

Ribbon:	Raster Tools > Edit drop-down > Merge Vector
Command:	IVMERGE

The **Merge Vector** tool helps in merging vectors into an existing image or a new raster image. To do so, select the **Merge Vector** tool from the **Edit** drop-down of the **Raster Tools** tab; the cursor will change into a crosshair and you will be prompted to select objects. Select the desired vector objects and press ENTER to start the vector merge operation.

TUTORIALS

Before starting the tutorial, you need to download and save the tutorial files on your computer. To do so, follow the steps given below.

1. Download the *c03_rd_2016_tut.zip* file from *http://www.cadcim.com*. The path of the file is as follows:
   Textbooks > Civil/GIS > AutoCAD Raster Design > Exploring AutoCAD Raster Design 2016.

2. Now, save and extract the downloaded folder at the following location:

   C:\AutoCADRasterDesign2016

Notice that *c03_rd_2016_tut* folder is created within the *AutoCADRasterDesign2016* folder.

Tutorial 1   Inverting and Image Cropping

In this tutorial, you will start the AutoCAD Raster Design application and then insert a bitonal image using the **Insert** tool. Next, you will crop the bitonal image by using the marked points on the image and then invert the given image so that the image background changes to the AutoCAD background. Figure 3-9 shows the raster image to be cropped and inverted.

(Expected time: 30 min)

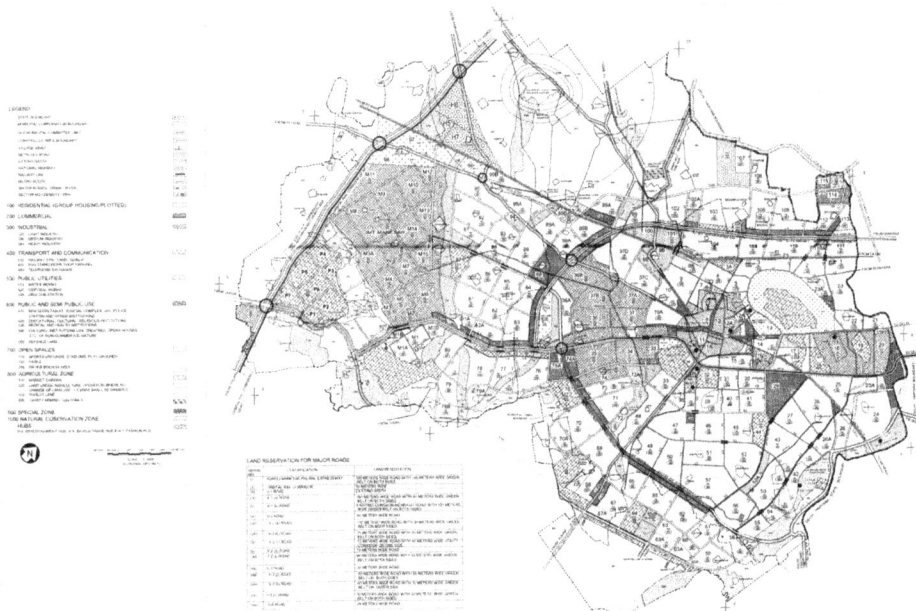

Figure 3-9 *The bitonal master plan map*

The following steps are required to complete this tutorial:

a. Start AutoCAD Raster Design application and open a new template.
b. Insert the bitonal image.
c. Invert the image
d. Crop the bitonal image by using marked points on the image.
e. Save the drawing file.

Starting AutoCAD Raster Design and Opening the New Template

1. Start AutoCAD Raster Design application and choose the **New** from the Application Menu; the **Select Template** dialog box is displayed.

2. Select the **acad.dwt** template and choose the **Open** button to open the template file.

Inserting the Bitonal Image

1. Choose the **Insert** tool from the **Insert & Write** panel of the **Raster Tools** tab; the **Insert Image** dialog box is displayed.

2. In this dialog box, select the **Quick Insert** radio button from the **Insert Options** area.

3. Browse to the location *C:\AutoCADRasterDesign2016\c03_rd_2016_tut\c03_tut01* and then select the **Master_plan.bmp** file.

4. Choose the **Open** button; the **Insert Image** dialog box is closed and the image is inserted in the drawing.

5. Enter **Z** at the Command prompt area and then press ENTER; you are prompted to specify the zoom option.

6. Enter **E** and then press ENTER; the drawing zooms to its extents.

> **Note**
> *You can zoom the drawing to its extents by using the* **Zoom to Extents** *tool from the* **Manage** *&* **View** *panel of the* **Raster Tools** *tab.*

Inverting the image

1. Choose the **Invert** tool from the **Cleanup** drop-down in the **Edit** panel of the **Raster Tools** tab; the background color of the image turns black, as shown in Figure 3-10.

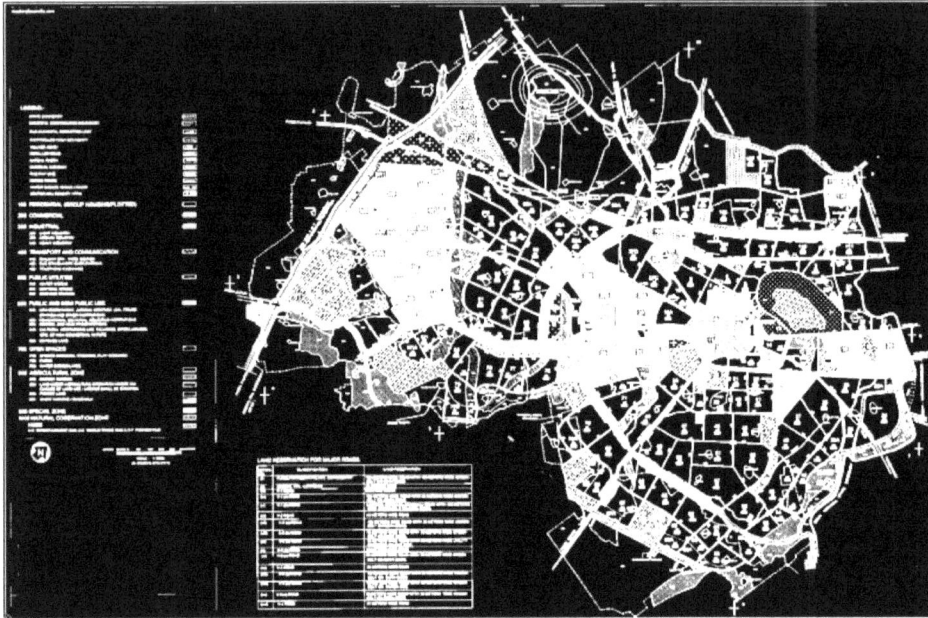

Figure 3-10 *The inverted image*

Cropping the Bitonal Image

1. Choose the **Polygonal Region** tool from the **Crop** drop-down in the **Edit** panel of the **Raster Tools** tab; you are prompted to specify the first point on the image.

2. Zoom in and click at the first point (1) on the image, refer to Figure 3-11. Notice that the image area to be cropped is already marked on the image.

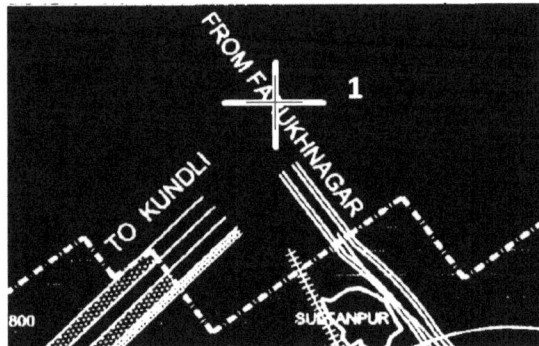

Figure 3-11 *The first point on the image*

3. Click on the center of all the 10 specified points on the image in the drawing area.

4. After clicking on all the points on the image, right-click on the drawing; a shortcut menu is displayed.

5. Choose the **Enter** option from the menu; the image is cropped and the frame size automatically compensates to the image border, as shown in Figure 3-12.

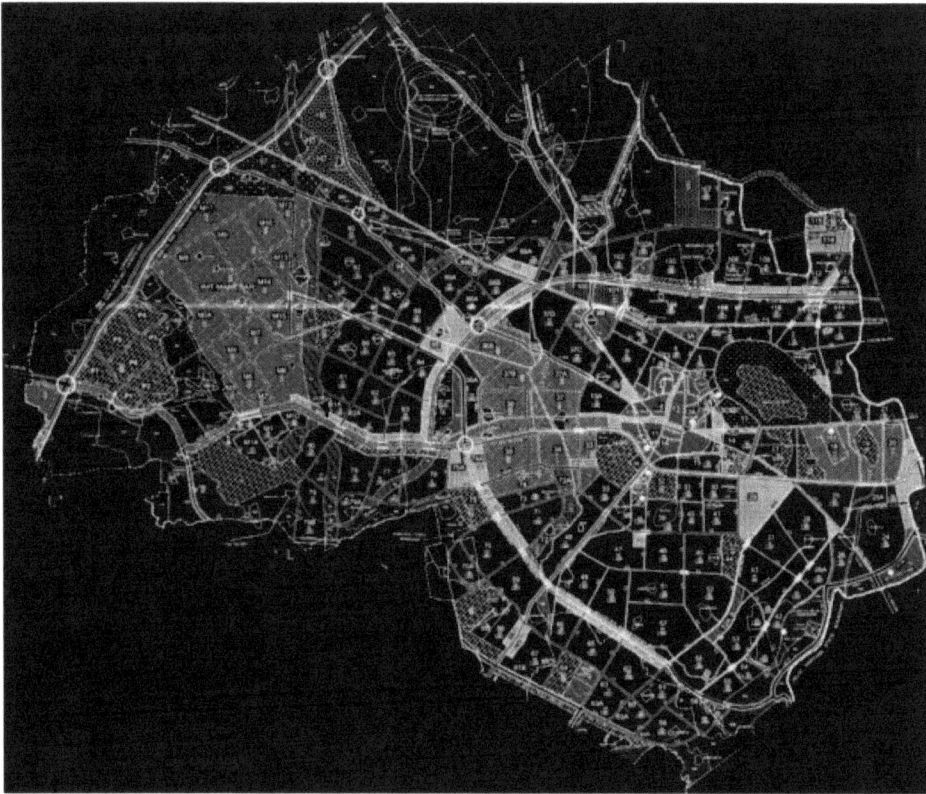

Figure 3-12 *The cropped image*

Note
You can also press ENTER to end the cropping process.

Saving the File

1. Choose the **Save As** option from the Application Menu; the **Save Drawing As** dialog box is displayed.

2. In this dialog box, browse to the following location:

   C:\AutoCADRasterDesign2016\c03_rd_2016_tut\c03_tut01

3. In the **File name** edit box, enter **c03_Tut01_Result**.

4. Choose the **Save** button; the dialog box is closed and the drawing file is saved with the name **c03_Tut01_Result.dwg** at the specified location. Also, the **Save Image** dialog box is displayed, refer to Figure 3-13.

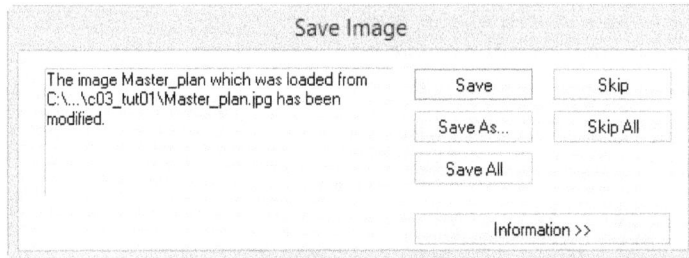

Figure 3-13   *The **Save Image** dialog box*

5.   Choose the **Save As** button in the **Save Image** dialog box; the **Save As** dialog box is displayed.

6.   In this dialog box, browse to the following location:

        C:\AutoCADRasterDesign2016\c03_rd_2016_tut\c03_tut01

7.   In the **File name** edit box, enter **Cropped_Master_Plan**.

8.   Ensure that the **Windows Bitmap (*.bmp, *.dib, *.rle)** option is selected from the **Files of type** drop-down list and then choose the **Save** button. The file is saved at the specified location.

Closing the File

1.   Choose the **Close** option from the Application Menu; the file is closed. Ignore the message box displayed, if any.

Tutorial 2   Despeckle and Image Touchup

In this tutorial, you will use the **Despeckle** tool to cleanup raster image and to remove scanning artifacts from the image. Next, you will use the **Touchup** tool to correct imperfections in the scanned bitonal images.   **(Expected time: 30 min)**

The following steps are required to complete this tutorial:

a.   Start AutoCAD Raster Design application and open the *c03_Tut02.dwg* file.
b.   Despeckle the bitonal image.
c.   Image touchup.
d.   Save the drawing file.

Starting AutoCAD Raster Design application and opening the Drawing File

1. Choose the **Open** button from the **Quick Access** toolbar; the **Select File** dialog box is displayed.

2. In this dialog box, browse to the following location:

    C:\AutoCADRasterDesign2016\c03_rd_2016_tut\c03_tut02

3. Select the **c03_Tut02.dwg** file and choose the **Open** button; the selected drawing is opened and the geometry is displayed in the drawing window.

4. Enter the **IDEPTH** command in the Command prompt area, and press ENTER.

5. Select the **Bitonal** option from the Command prompt area, the image is converted into bitonal image, as shown in Figure 3-14.

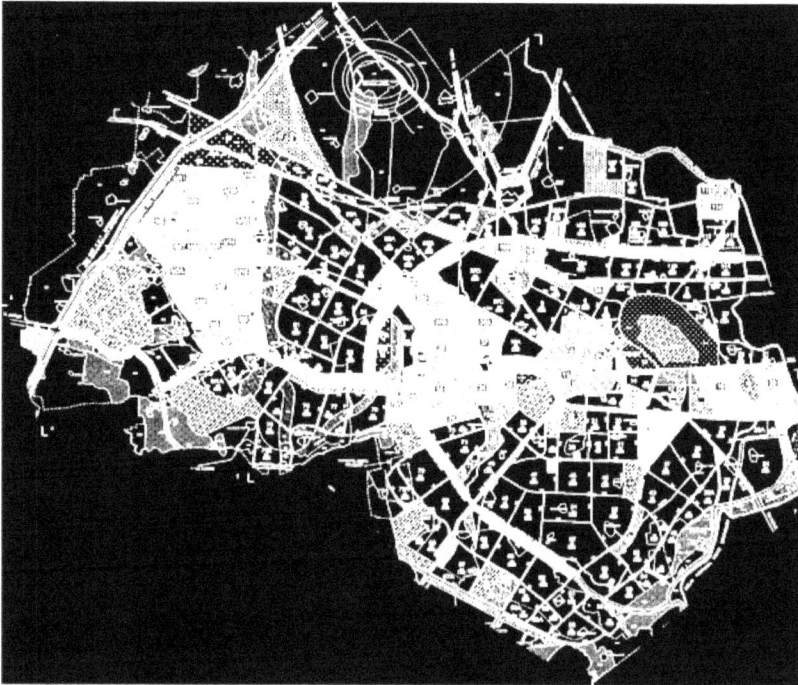

Figure 3-14 *The image converted to bitonal image*

Despeckling the Image

In this section, you will remove the speckle from the bitonal image.

1. Choose the **Despeckle** tool from the **Cleanup** drop-down of the **Raster Tools** tab.

2. Zoom in on the desired location in the drawing to locate the speckle on the image, refer to Figure 3-15.

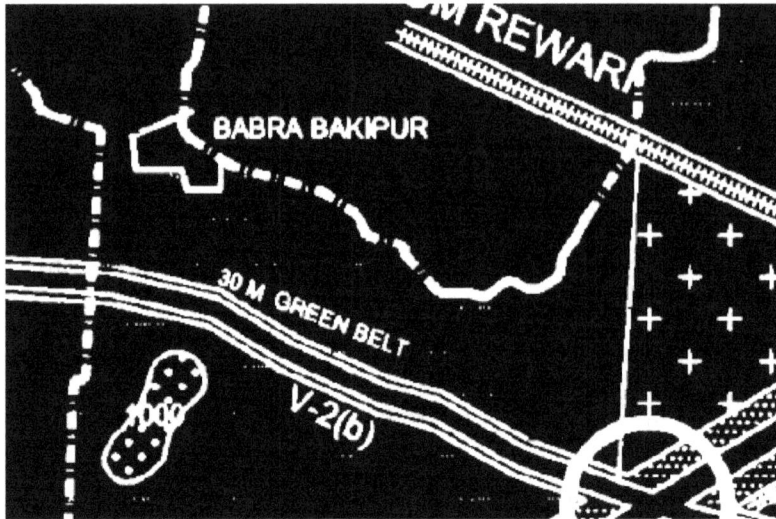

Figure 3-15 Zoomed image with speckles

3. Press the ENTER button or select the **entire Image** option from the Command prompt area.

4. Now, pick a speckle to set the size of the speckle; color of the speckles in the image changes to red.

5. Next, press the ENTER button to delete the selected speckles from the entire image.

6. Press the ESC key to terminate the command.

Note
*You can also use the **Window** or **Polygon** option to deselect speckles inside a rectangle or polygon.*

Performing the Image Touchup

1. Choose the **Touchup** tool from the **Edit** panel of the **Raster Tools** tab; the **Touchup** toolbar is displayed.

2. Next, zoom in the edge that you want to correct as shown in Figure 3-16.

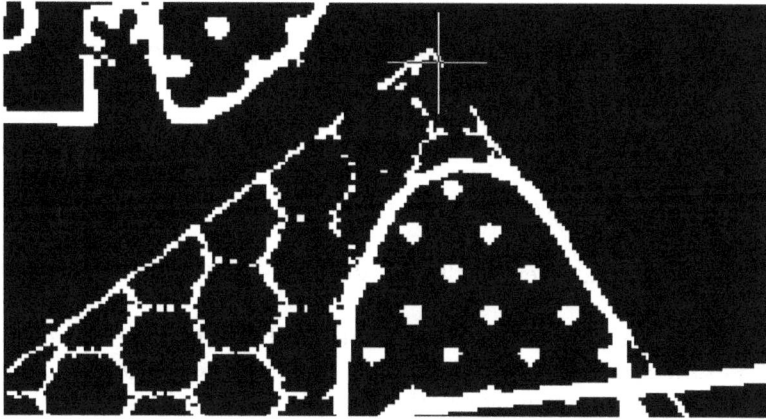

Figure 3-16 *Edge of the line to be corrected*

3. Choose the **Diagonal brush** tool from the **Touchup** toolbar and then choose the **Resize brush** tool; the cursor changes into a crosshair.

4. Click on the edge of the line that you want to repair and then specify another point along the same edge. Now click on another point to specify a brush width that is approximately equal to the width of the line; a drawing brush of the specified size is created.

5. Draw a straight line with the brush by clicking on the gap area; the gap is filled as shown in Figure 3-17.

Figure 3-17 *Lines corrected by using the **Touchup** tool*

6. Press ENTER to terminate the command.

Saving the File

1. Choose the **Save As** option from the Application Menu; the **Save Drawing As** dialog box is displayed.

2. In this dialog box, browse to the following location:

    C:\AutoCADRasterDesign2016\c03_rd_2016_tut\c03_tut02

3. In the **File name** edit box, enter **c03_Tut02_Result**.

4. Choose the **Save** button; the dialog box is closed and drawing file is saved with the name **c03_Tut02_Result.dwg** at the specified location. Also, the **Save Image** dialog box is displayed, as shown in Figure 3-18, informing you that the raster in the drawing has been modified.

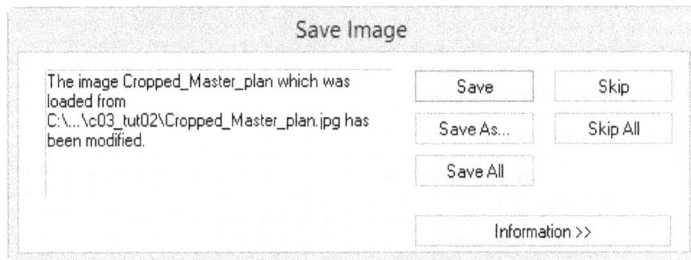

Figure 3-18 The Save Image dialog box

5. Choose the **Save As** button from the dialog box; the **Save As** dialog box is displayed.

6. In this dialog box, browse to the following location:

    C:\AutoCADRasterDesign2016\c03_rd_2016_tut\c03_tut02

7. In the **File name** edit box, enter **Touchup_Master_Plan**.

8. Ensure that the **Windows Bitmap (*.bmp, *.dib, *.rle)** option is selected from the **Files of type** drop-down list and then choose the **Save** button; the image is saved.

Closing the File

1. Choose the **Close** option from the Application Menu; the file is closed. Ignore the message box displayed, if any.

Tutorial 3   Image Merge

In this tutorial, you will merge raster images and save them using the **Merge Images** tool.
(**Expected time: 30 min**)

The following steps are required to complete this tutorial:

a. Start AutoCAD Raster Design application and open the *c03_Tut03.dwg* file.
b. Merge raster images.
c. Save the drawing file.

Opening the Drawing File

1. Choose the **Open** button from the **Quick Access** toolbar; the **Select File** dialog box is displayed.

2. In this dialog box, browse to the following location:

    C:\AutoCADRasterDesign2016\c03_rd_2016_tut\c03_tut03

3. Next, select the **c03_Tut03.dwg** file and choose the **Open** button; the selected drawing is opened in the drawing window.

Merging Raster Images

1. Select the **Merge Images** tool from the **Edit** drop-down of the **Raster Tools** tab.

2. Select the images, as shown in Figure 3-19; and press ENTER.

Figure 3-19 Images to be selected

Tip
You can select the raster images by clicking on their frame or by creating a selection window.

3. Next, select the destination image, as shown in Figure 3-20; you will be prompted to erase source images.

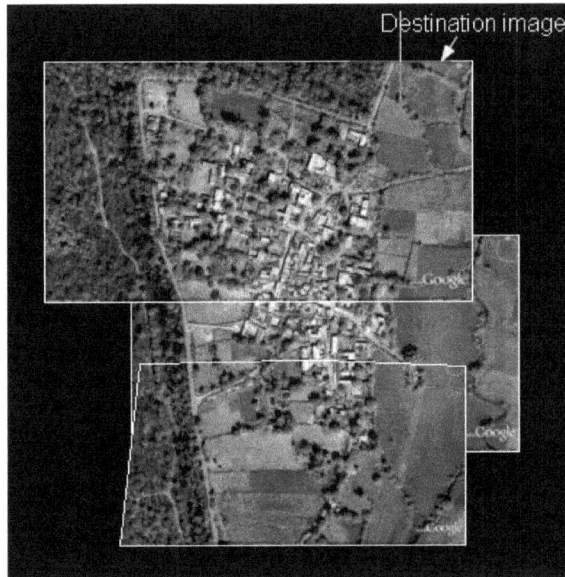

Figure 3-20 Showing destination image

4. Select the **Yes** option from the Command prompt area; the **AutoCAD Raster Design** message box will be displayed.

5. Choose the **Yes** button from the **AutoCAD Raster Design** message box to detach the source images from the destination image; once again the **AutoCAD Raster Design** message box will be displayed.

6. Choose the **Yes** button; the dialog box will be closed and the images will be merged.

Saving the File

1. Choose **Save As** from the Application Menu; the **Save Drawing As** dialog box is displayed.

2. In this dialog box, browse to the following location:

    C:\AutoCADRasterDesign2016\c03_rd_2016_tut\c03_tut03

3. In the **File name** edit box, enter **c03_Tut03_Result**.

4. Choose the **Save** button; the dialog box is closed and drawing file is saved with the name **c03_Tut03_Result.dwg** at the specified location. Also, the **Save Image** dialog box is displayed.

5. In the **Save Image** dialog box, choose the **Save** button; the **Save As** dialog box is displayed.

6. In this dialog box, browse to the following location:

    C:\AutoCADRasterDesign2016\c03_rd_2016_tut\c03_tut03

7. In the **File name** edit box, enter **Merge_image**.

8. Ensure that the **JPEG File Interchange Format (*.jpg, *.jpeg)** option is selected from the **Files of type** drop-down list and then choose the **Save** button; the **Save As** dialog box is closed and the **Encoding Method** dialog box is displayed.

9. Choose the **Maximum Quality** option from the **Encoding** list in this dialog box and then choose the **Finish** button; the raster image is exported to the specified location.

Closing the File

1. Choose the **Close** option from the Application Menu; the file is closed. Ignore the message box displayed, if any.

Self-Evaluation Test

Answer the following questions and then compare them to those given at the end of this chapter:

1. Which of the following commands is used to convert a raster image into bitonal image?

    (a) **IDELETE**   (b) **IDEPTH**
    (c) **IDESKEW**   (d) **IBIAS**

2. Which of the following options is used to crop a defined polygonal region?

    (a) **Polygonal Region**   (b) **Circular Region**
    (c) **Diagonal Region**   (d) All of the above

3. Which of the following options is used to remove speckle from an image?

    (a) **Despeckle**   (b) **Bias**
    (c) **Deskew**   (d) All of the above

4. You can use the _____ tool from the **Touchup** toolbar to edit any existing line on an image.

5. You can change the direction of the brush by clicking on the **Resize brush** tool from the **Touchup** toolbar. (T/F)

6. You can use the **Line Entity** tool to crop an image. (T/F)

7. You can mirror or flip an image along either horizontal or vertical axis using the **Mirror** tool. (T/F)

8. You can remove more than one raster line segment at a time by using the **Rub Polyline** command. (T/F)

Review Questions

Answer the following questions:

1. The **Raster Snap** makes it easy to select points on raster entities when you are using the _____ tools.

2. You can use the _____ filter to reverse the light and dark shades in any image.

3. You can use the **Bias** tool to correct distortions in the _____ of an image.

4. You can use the **Merge Images** tool to remove all the raster data inside a defined polygonal region. (T/F)

5. In AutoCAD Raster Design, you can remove raster entities from bitonal images using the **Line**, **Arc**, and **Circle** tool. (T/F)

6. Snap mode determine where your cursor will snap to as you move it around an image. (T/F)

7. You can define the size of autosnap marker and aperture size in the **Raster Snap** options area. (T/F)

8. In AutoCAD Raster Design, if you choose more than one snap mode, then the crosshair snaps to the closest snap points. (T/F)

EXERCISE

Exercise 1

Download the *c03_rd_2016_exr.zip* from *http://www.cadcim.com* and convert the given image into a bitonal image. Next, crop the image by selecting the four corners of the image and remove speckles from the image.   **(Expected time: 40 min)**

Answers to Self-Evaluation Test
1. b, **2.** a, **3.** a, **4. Diagonal brush**, **5.** T, **6.** F, **7.** T, **8.** T

Chapter 4

Image Processing

Learning Objectives

After completing this chapter, you will be able to:
- *Understand image processing*
- *Understand histogram of an image*
- *Filter an image*

INTRODUCTION

Sometimes raster images can be difficult to interpret. AutoCAD Raster Design is the only product of AutoCAD which allows the user to perform image processing. AutoCAD Raster Design provides tools to archive and update the images for better understanding. Image processing techniques have been developed to aid the interpretation of satellite images and to extract as much information as possible from the images. In this chapter, you will learn various image processing techniques.

IMAGE PROCESSING

AutoCAD Raster Design provides tools to process a scanned image to get an enhanced image. This software is widely used in research area as well as in engineering and computer science discipline. Basically in image processing, an optical image is imported into software to analyze and manipulate its quality. Image processing is a step by step process to enhance the quality of the image. The major processes involved in image processing are given below:

1. Visualize image to observe its features
2. Sharpen the image
3. Remove noise and retrieve original information
4. Recognize pattern

Image processing can be classified into two types: analog and digital. The analog or visual image processing technique is used for processing hard copy images and photographs. This technique depends on how the images are interpreted.

Digital image processing is a computer based technique and it has a various phases to enhance the quality of the image. Mainly three steps are required to be performed in this technique: image pre-processing, image enhancement and transformation, and image analysis.

In AutoCAD Raster Design, various image processing tools are used to enhance the appearance of an image. These tools change the image pixels permanently to enhance the quality of the image. Various image processing tools are described in Table 4-1.

Table 4-1 *Image processing tools and their description*

Histogram	Used to change the brightness and contrast of an image.
Convolve	Used to smoothen and sharpen an image.
Bitonal Filters	Used to cleanup bitonal image.
Change Density	Used to change the density or resolution of a raster image.
Change Color Depth	Used to change the radiometric resolution.
Palette Manager	Used to manipulate the color of an image by using this palette.

Histogram

Ribbon: Raster Tools > Edit > Process Image > Histogram
Command: IHISTOGRAM

AutoCAD Raster Design provides tools to permanently change the appearance of grayscale or color image. Histogram is a bar graph which represents the number of pixels per pixel shade in a selected image. You can change the effect of brightness/contrast, equalizing, color to grayscale or tonal adjustments on a portion of the given image or on the entire image.

For changing the image appearance, insert the image into your drawing. Next, choose the **Histogram** tool from the **Process Image** drop-down list of the **Raster Tools** tab; you will be prompted to specify the first corner point of selection window. You can also select the **entire Image**, **Clip region**, **Polygon**, or **Existing** option from the Command prompt area to specify the region in the image. Click on the image to specify the first corner point of selection window. Similarly, specify the second corner point which is diagonally opposite to the first point; the **Histogram** dialog box will be displayed with the **Tonal Adjustment** tab chosen, as shown in Figure 4-1. The various options in this dialog box are discussed next.

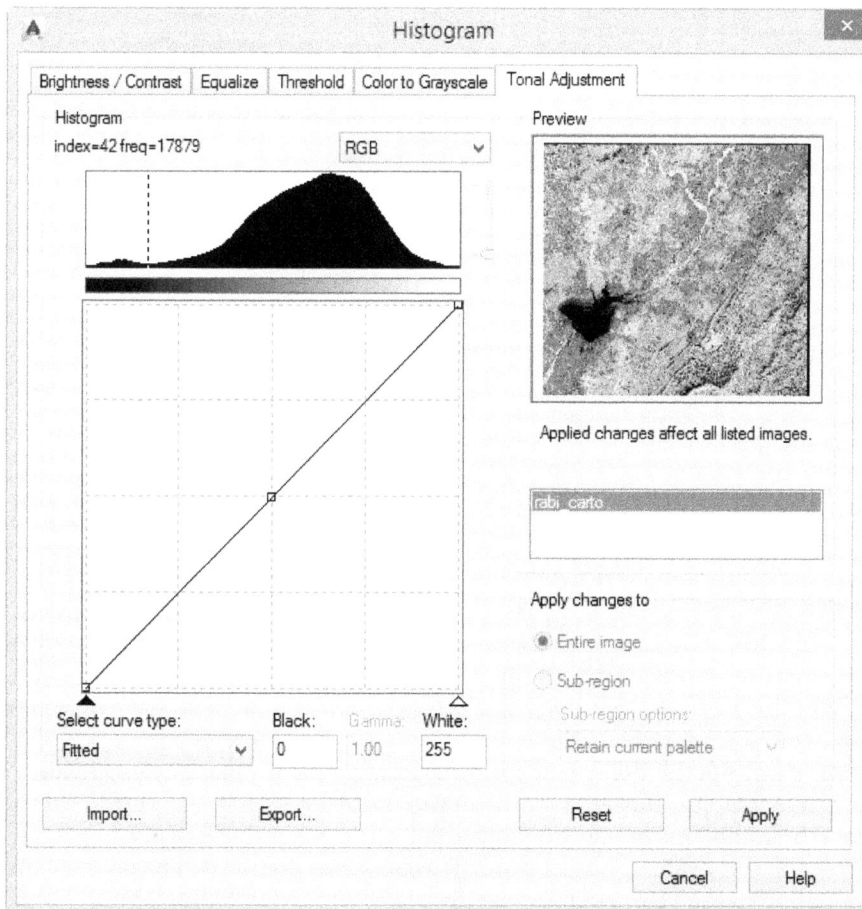

Figure 4-1 *Various options in the **Histogram** dialog box*

The Tonal Adjustment Tab

In the **Histogram** dialog box, the **Tonal Adjustment** tab is chosen by default. In the **Histogram** area, you can select the **Red**, **Green**, **Blue**, or **RGB** option from the **RGB** drop-down list. These options will not be enabled if you insert a bitonal image into your drawing. By modifying the contrast curve, the histogram graph of the image will be updated automatically, as shown in Figure 4-2.

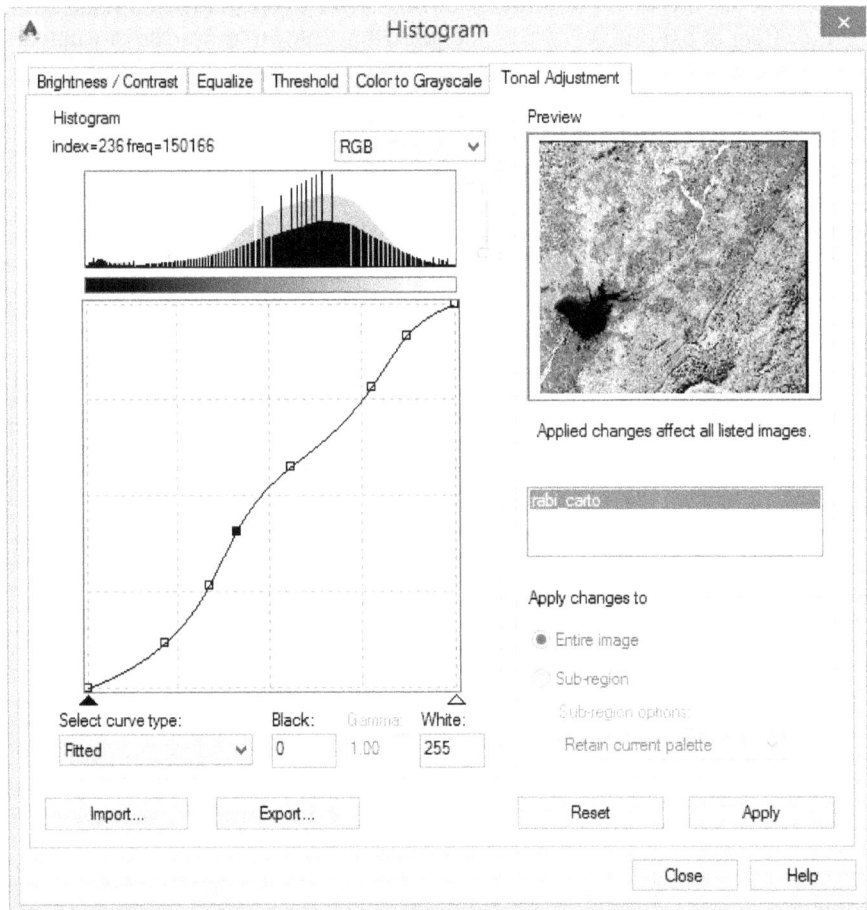

Figure 4-2 The updated histogram graph with change in the contrast curve

You can define the contrast curve representation by selecting the required option from the **Select curve type** drop-down list. The **Gamma** option results in an exponential curve, the **Fitted** option results in a user-defined smooth curve, and the **Piecewise Linear** option results in a user defined straight line instead of a curve.

You can specify a value in the **Black** edit box or you can adjust the black colored slider just below the contrast curve to adjust the color values in the image. Similarly, you can specify a required value in the **White** edit box or you can adjust the white colored slider just below the contrast curve to adjust the color values in the image. The **Gamma** edit box will be enabled if you select the **Gamma** option in the **Select curve type** drop-down list. The value in the **Gamma** edit box must be in the range of 0.10 to 10.00. The **Preview** area of the **Histogram** dialog box shows a preview of the selected image, refer to Figure 4-2.

Note

*In the **Tonal Adjustment** tab, the **Apply changes to** area will be activated only if any option other than the **entire Image** option is selected while specifying the image region for histogram equalization in the Command prompt area.*

The **Apply** button is used to fix the specified changes in the entire image or sub-region.

The Color to Grayscale Tab

The options in the **Color to Grayscale** tab, refer to Figure 4-3, are used to convert a color image into an 8-bit grayscale image and reduce the size of the image. The **Description** area provides the details of the functions being performed.

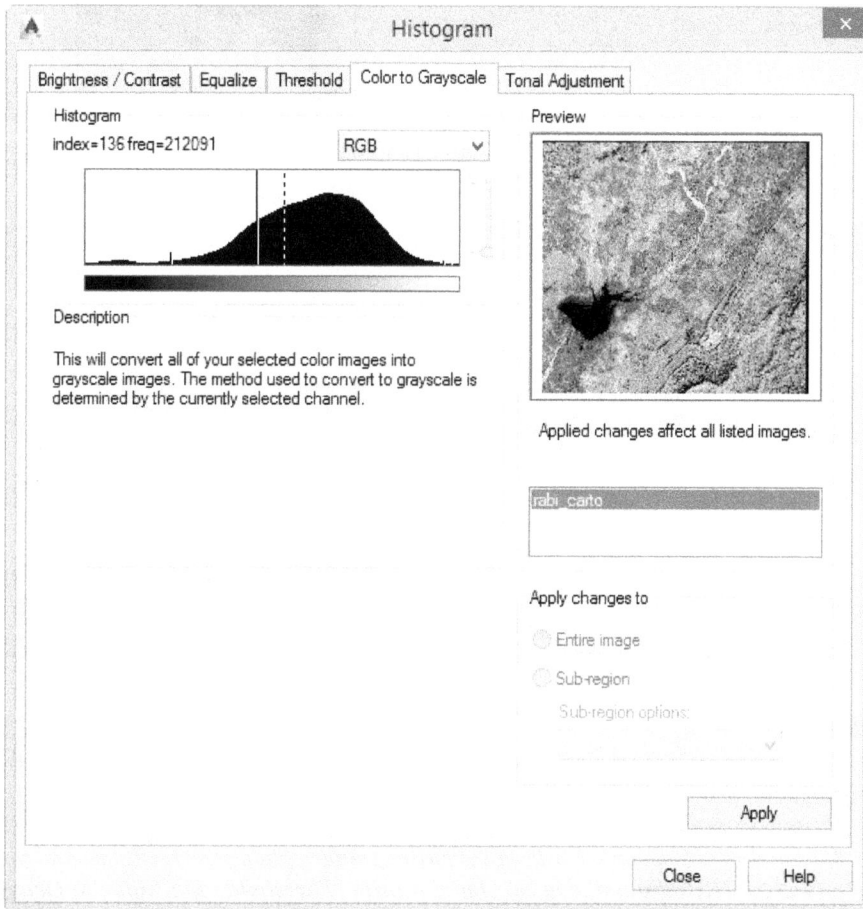

*Figure 4-3   The **Histogram** dialog box with the **Color to Grayscale** tab chosen*

The Threshold Tab

The options in the **Threshold** tab are used to convert a color image into a binary image. In this tab, a slider is available in the **Threshold** area. If you move this slider towards the left, the pixels that are represented at the left in the histogram graph will turn black and the pixels that are represented at the right will turn white.

The **Preview** area in this tab also displays the modification made in the image by adjusting the slider in the **Threshold** area, refer to Figure 4-4.

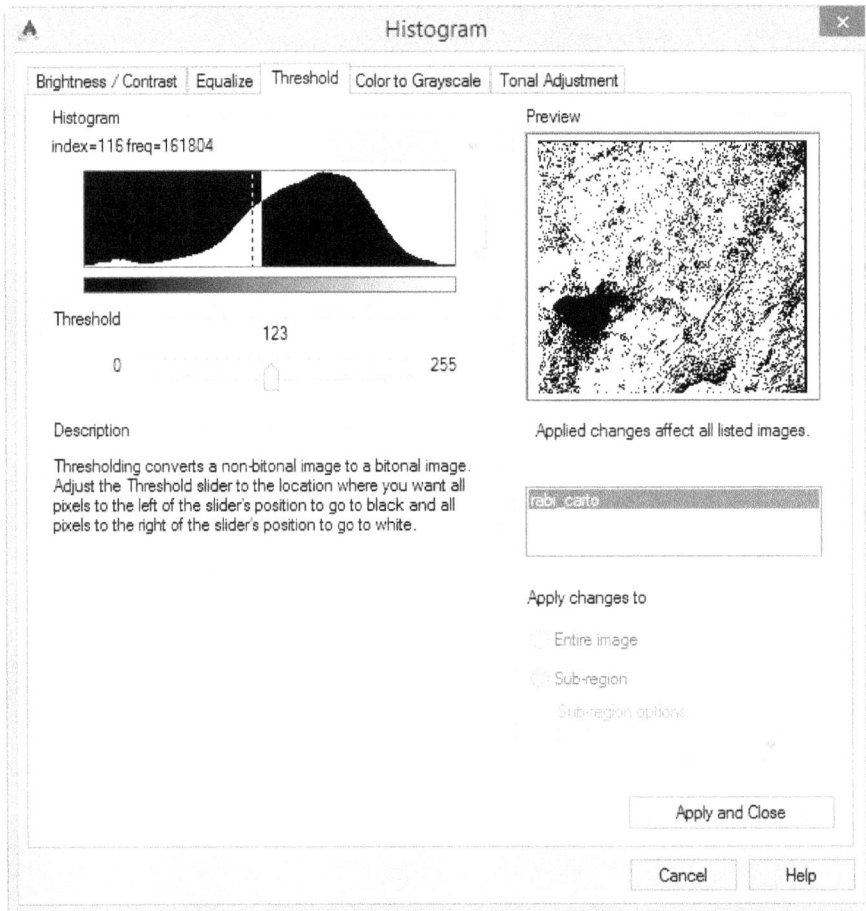

*Figure 4-4 The **Histogram** dialog box with the **Threshold** tab chosen*

Note

*1. If you choose the **Threshold** tab, the **RGB** channel drop-down list will be disabled in the **Histogram** area of the **Histogram** dialog box.*

*2. If you select any option other than the **entire Image** option from the Command prompt area to invoke the **Histogram** dialog box, the **Equalize, Threshold,** and **Color to Grayscale** tabs in the dialog box will remain disabled. Also, the **Equalize** tab will be activated only if you make any changes in the **Brightness/Contrast** tab or in the **Tonal Adjustment** tab.*

The Equalize Tab

The options in the **Equalize** tab are used to maximize the details of a given image. By using the options in this tab, the maximum percentage of pixels containing approximately same color are equalized. The darkest pixels are changed to black and the lightest pixels are changed to white. The remaining pixels are converted to colors in between black and white.

When you equalize an image, you can select which channel to equalize from the **RGB** channel drop-down list and then by adjusting the vertical slider on the histogram graph.

> **Tip**
> *When you adjust the vertical slider in the **Histogram** area of the **Equalize** tab, the index and frequency values will be updated automatically. Index displays the intensity of each shade in the histogram between 0 to 255 and frequency shows the number of pixels of the corresponding index value.*

In the **Preview** area, the preview of the image will be displayed according to the modifications made in the image, refer to Figure 4-5.

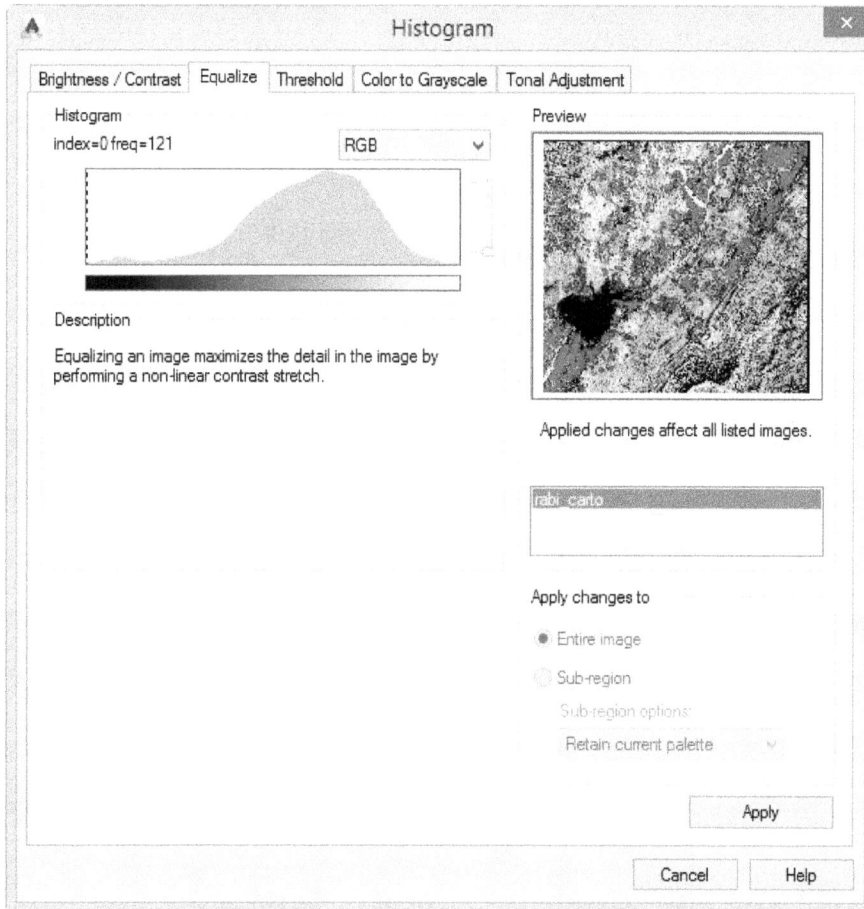

*Figure 4-5 The **Histogram** dialog box with preview in the **Equalize** tab*

The Brightness/Contrast Tab

The options in the **Brightness/Contrast** tab are used to adjust the brightness and contrast settings of the color and grayscale images.

You can adjust the sliders in the **Brightness** and **Contrast** areas to increase or decrease the brightness and contrast values in the image. Select the **Apply** button to update the image, as shown in Figure 4-6.

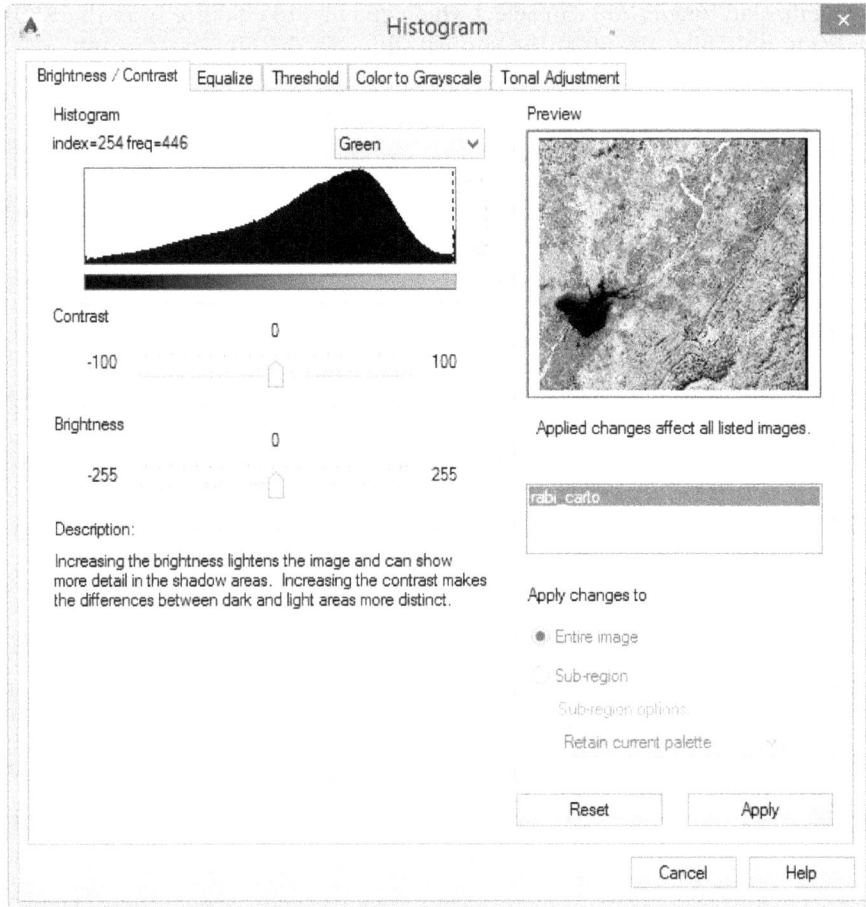

*Figure 4-6 The **Histogram** dialog box with the **Brightness/Contrast** tab chosen*

Convolve/Filtering

Ribbon: Raster Tools > Edit > Process Image > Convolve
Command: ICONVOLVE

The process of applying filters to an image is called convolution. A filter is a kernel (a small array of pixels) applied to each pixel and its neighbors within an image. The filtering process in AutoCAD Raster Design uses smoothing filters and sharpening filters to remove the unnecessary details from the original image. You can use lowpass and median filters as smoothing filters to smoothen the given image. To sharpen the given image, you can use highpass filters as sharpening filters.

To enhance a grayscale image, choose the **Convolve** tool from the **Process Image** drop-down; the **Image Filters** dialog box will be displayed, as shown in Figure 4-7.

Note
Filters can be applied only on grayscale images. If you insert an image other than the grayscale image, an error message will be displayed stating that convolve filters will only work on grayscale images.

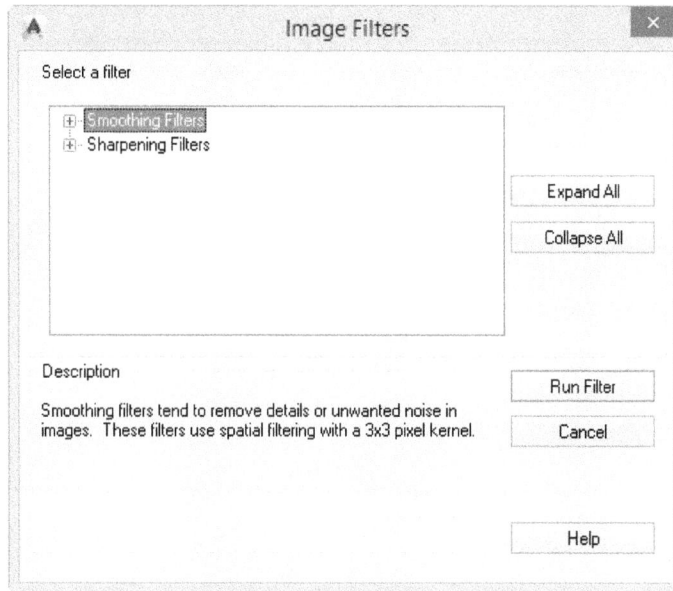

Figure 4-7 *The* *Image Filters* *dialog box*

You can choose any filter for sharpening or smoothing the image from the **Select a filter** area of the **Image Filters** dialog box. In the **Description** area, details of the selected filter will be displayed.

Smoothing Filters
Smoothing in AutoCAD Raster Design is a procedure to reduce noise from an image. It gives a less pixelated image.

Expand the **Smoothing Filters** node in the list box available in the **Select a filter** area; the **Lowpass Filtering** and **Median Filter** nodes will be displayed. These filters are discussed next.

Lowpass Filtering
The lowpass filters sharpens the high frequency pixels into low frequency pixels to lessen the severity of tone change. The **Lowpass Filtering** node contains four sub-nodes: **Lowpass Filter #1**, **Lowpass Filter #2**, **Lowpass Filter #3**, and **Lowpass Filter #4**. Brief description of these sub-nodes is given next.

The **Lowpass Filter #1** sub-node differences between target pixel value and its side to side and top to bottom neighbors and takes out its average. While calculating the average, the diagonally adjacent pixels are ignored. To run this filter, choose the **Lowpass Filter #1** by expanding **Smoothing Filters > Lowpass Filtering** nodes, refer to Figure 4-8. Next, choose the **Run Filter** button from the **Image Filters** dialog box.

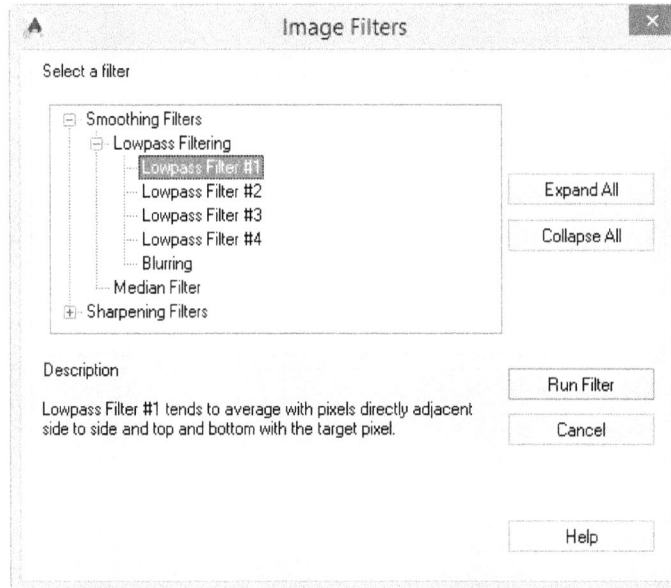

Figure 4-8 *The **Image Filters** dialog box with the **Lowpass Filter #1** sub-node selected*

The **Lowpass Filter #2** sub-node is used to average all the pixels surrounding the high frequency pixel.

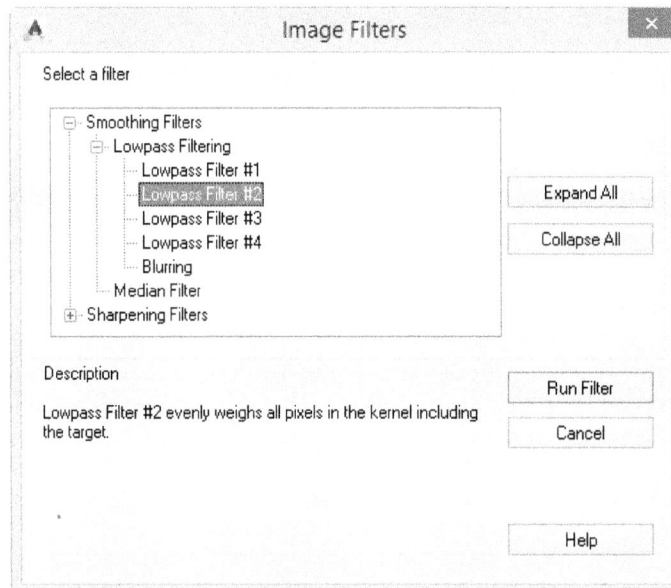

Figure 4-9 *The **Image Filters** dialog box with the **Lowpass Filter #2** sub-node selected*

The **Lowpass Filter #3** sub-node gives less blurred image by providing extra weight on the target pixel and its surrounding pixels.

The **Lowpass Filter #4** sub-node targets side to side and top to bottom pixels more heavily than those that are diagonally connected to the specified pixel. It results in less blurred image than the other filters.

The **Blurring** sub-node results in slightly brighter image by weighing all pixels in the kernel including target pixel. This filter is similar to **Lowpass Filter #2**.

Median Filter

Median filtering is a non-linear filtering technique. Median filters are used to remove impetuous noise from the image. It also removes spiky pixels and distributes bright pixels around the image. It is much better at preserving edges while reducing random noise from the image. The median filter is relatively complex to compute. To find the median value, it is necessary to find the average value of all neighboring pixels. To use this filter, choose the **Median Filter** node in the **Smoothing Filters** node, refer to Figure 4-10 and then choose the **Run Filter** button in the **Image Filters** dialog box; the **Median Filter** dialog box will be displayed. Next, enter the filter size in the **Filter size (odd integer 3-25)** edit box and choose **OK**.

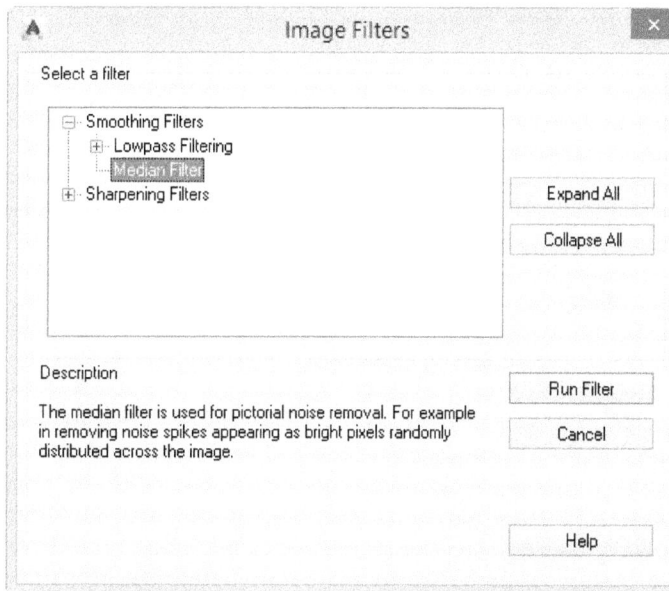

Figure 4-10 *The* *Image Filters* *dialog box with the* *Median Filter* *sub-node selected*

Sharpening Filters

In AutoCAD Raster Design, you can sharpen the given image by using image sharpening filters such as highpass and edge detection filters. Image sharpening refers to any enhancement technique that highlights the edges and fine details in an image. It is used to increase the local contrast and therefore sharpens the details of the image. The key point in the effective sharpening process lies in the choice of the highpass filtering operation. Highpass filters are linear filters and provide more accurate images by removing the noise. Different types of the sharpening filters are discussed next in detail.

Highpass Filters

To use highpass filters, expand **Sharpening Filters** node from the **Image Filters** dialog box, as shown in Figure 4-11.

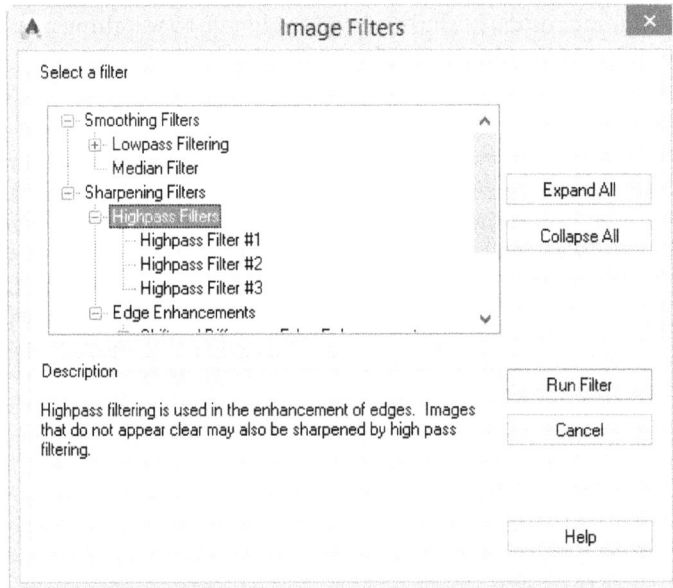

*Figure 4-11 The **Image Filters** dialog box with the **Highpass Filters** sub-node selected*

If you expand the **Highpass Filters** tree, you will notice three major highpass filters which are discussed next.

The **Highpass Filter #1** sub-node averages all surrounding pixels to the specified pixel but ignores diagonally connected pixels. It results in more sharp image due to the rise of frequency value to a higher value.

The **Highpass Filter #2** sub-node increases the low frequency value into high frequency in the targeted pixel.

The **Highpass Filter #3** sub-node concentrates more heavily on the side to side and top to bottom pixels than the diagonal pixels.

Edge Enhancements

AutoCAD Raster Design provides various filters to enhance the edges of the image by using object extraction and object recognition techniques. These filters are discussed next.

The **Shift and Difference Edge Enhancements** filter shifts image edges by one pixel and subtracts them from original image. AutoCAD Raster Design provides three different types of **Shift and Difference Edge Enhancements** filter: vertical, horizontal, and vertical and horizontal. The **Vertical** filter enhances the vertical edges from left to right. Similarly, the **Horizontal** filter enhances horizontal edges from top to bottom. The **Vertical and Horizontal** filter combines both vertical and horizontal edge enhancements from top to bottom and left to right.

The **Laplace Edge Enhancements** filter highlights all edges in the image. This filter can be calculated using standard convolution methods. This filter enhances the image in all directions. The **Laplace #1** filter uses the target pixel and its surrounding neighbors. The **Laplace #2** filter applies high weightage values on its target to enhance the image quality. The **Laplace #3** filter assigns highest kernel value on the target pixel. Similarly, the **Laplace #4** filter adjusts horizontal and vertical pixels rather than the surrounding pixels.

In AutoCAD Raster Design, eight **Gradient Directional Edge Enhancements** filters are provided to enhance the grayscale image. These are: **North** filter, **Northeast** filter, **East** filter, **Southeast** filter, **South** filter, **Southwest** filter, **West** filter, and **Northwest** filter. Applying these filters results in a black background with image containing white outlines, as shown in Figure 4-12.

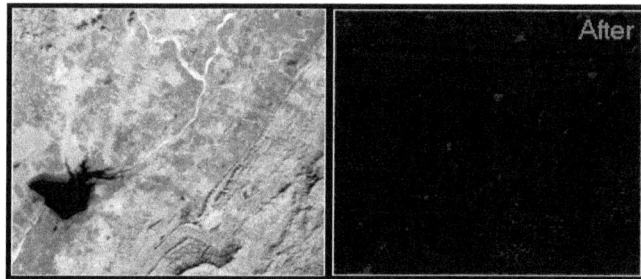

Figure 4-12 *Before and after applying the* **North** *directional edge enhancement filter*

The **Matched Filter Edge Enhancements** filters can be used in line destriping. These filters remove noise from the grayscale image in horizontal and vertical directions.

Bitonal Filters

Ribbon: Raster Tools > Edit > Process Image > Bitonal Filters
Command: IBFILTER

In AutoCAD Raster Design, you can enhance bitonal images by using the **Bitonal Filters** tool. This tool is used to remove speckles, deskew an image, and cleanup an image, as discussed in Chapter 3. You can use these bitonal filters on entire image, or on clipped region, polygon region, or existing view.

To enhance bitonal images, choose the **Bitonal Filters** tool from the **Process Image** drop-down of the **Edit** panel of the **Raster Tools** tab. Then, choose the desired option from the Command prompt area; the **Bitonal Filters** dialog box will be displayed, as shown in Figure 4-13.

In this dialog box, you can choose the desired option from the **Filter type** drop-down list. Some details about these filter types are given next.

The **Smooth** option removes unnecessary speckles and fill up breaks in raster lines. The **Thin** option trims raster entities by one pixel in the desired direction. The **Thicken** option thickens raster object edges by one pixel in the desired direction. The **Separate** option is used to separate thick lines from the image and the **Skeletonize** option is used to thicken raster lines in the desired location.

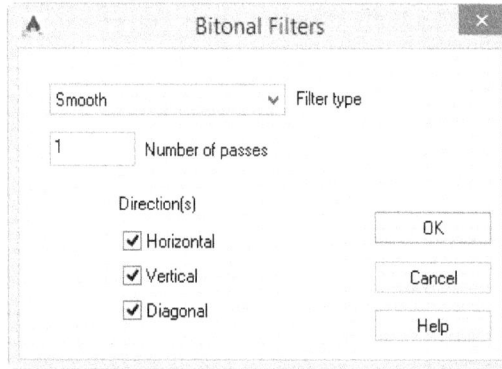

Figure 4-13 The **Bitonal Filters** *dialog box*

Note
*Different options in the **Bitonal Filters** dialog box are mostly used to edit or modify engineering drawings, manufacturing drawings, cadastral maps, and so on.*

In the **Number of passes** edit box, you can enter a value to specify the number of times the filter would run. If you want to run a filter more number of times, specify higher values in this edit box. This edit box will be inactive if you choose the **Skeletonize** option from the **Filter type** drop-down list.

From the **Direction(s)** area, you can select the direction in which the filter will be applied. By default, the **Horizontal**, **Vertical**, and **Diagonal** check boxes are selected which allows the filter to be applied in the horizontal, vertical, and diagonal directions, respectively.

Change Density

Ribbon:	Raster Tools > Edit > Process Image > Change Density
Command:	IDENSITY

 In AutoCAD Raster Design, the **Change Density** tool is used to increase or decrease the size of the original image. Choose the **Change Density** tool from the **Process Image** drop-down; the **Change Density** dialog box will be displayed. You can adjust the density of the multiple images using the options of the **Change Density** dialog box.

In the **Change Density** dialog box, the **Current Settings** area shows the details of the pixel density, pixel size, and dimensions of the image, as shown in Figure 4-14.

In the **New Settings** area, you can specify the new pixel density and pixel size. The image dimension will change according to the values specified by you. You can select the image resampling method from the **Resampling Method** drop-down list. The options in this drop-down list are discussed next.

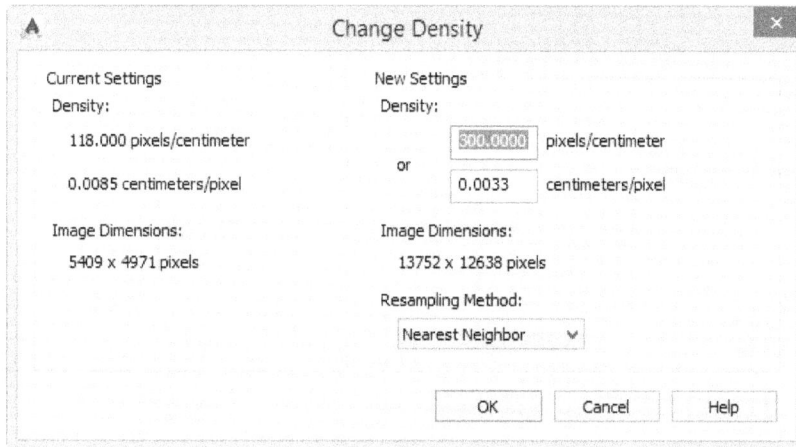

Figure 4-14 The **Change Density** *dialog box*

The **Bicubic** option resamples the image pixels in a 4*4 block of nearby pixels.
The **Bilinear** option resamples the image pixels in a 2*2 block of nearby pixels.
The **Lanczos** option resamples the image pixels in a 6*6 block of nearby pixels.
The **Mitchell** option resamples the image pixels using cubic spline technique.
The **Nearest Neighbour** option assigns new pixel values to its surrounding pixels and produces the crudest results.

Change Color Depth

Ribbon: Raster Tools > Edit > Process Image > Change Color Depth
Command: IDEPTH

Raster images are composed of pixels. These pixels store the bitonal information of the raster data. The information which is stored in each pixel is called depth and it is measured in bits. Higher the value of bitonal information, more colors will be stored in each pixel, whereas lesser the value of bitonal information, less colors will be stored. Note that an 8-bit image has 256 gray values and a 16 bit image has 65536 gray values.

In AutoCAD Raster Design, you can adjust the color depth of a raster image. To alter the depth of multispectral image, choose the **Change Color Depth** tool from the **Process Image** drop-down; you will be prompted to specify the new color type. Select the desired option from the Command prompt area; the color depth of the image will be adjusted. The following table shows the details of color type options provided in AutoCAD Raster Design.

Options	Color Depth	Gray Values
Bitonal	1-Bit	1 Color
Grayscale	8-Bit	256 Colors
Indexedcolor	8-Bit	256 Colors
True Color	24-Bit	16.7 Million Colors

Note
*The **Indexedcolor** option will be displayed in the Command prompt area when you insert a bitonal image.*

Palette Manager

Ribbon:	Raster Tools > Edit > Process Image > Palette Manager
Command:	IPAL

In AutoCAD Raster Design, you can alter individual colors of an 8-bit image using the **Palette Manager** dialog box. To invoke this dialog box, choose the **Palette Manager** tool from the **Process Image** drop-down in the **Raster Tools** tab. Figure 4-15 shows the **Palette Manager** dialog box.

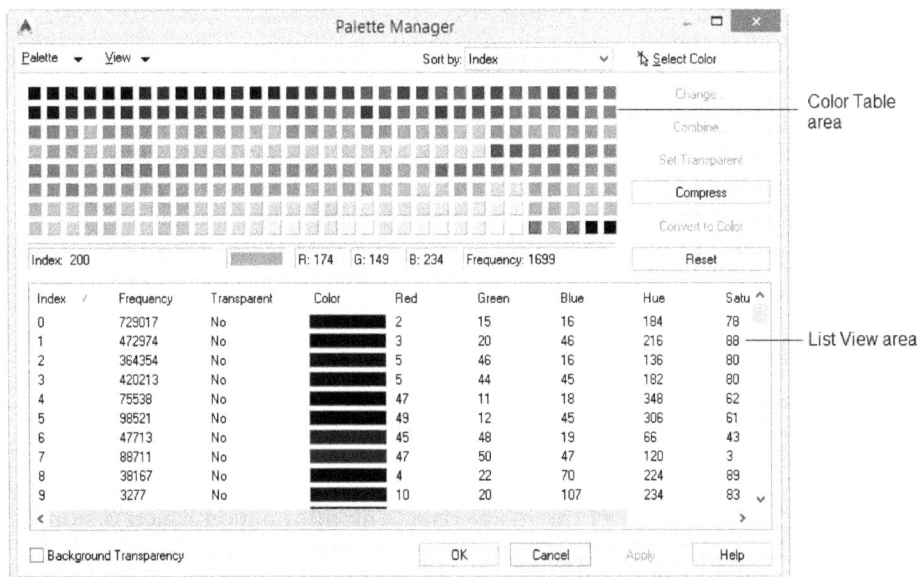

*Figure 4-15 The **Palette Manager** dialog box*

The **Palette Manager** dialog box provides detail information of entities. The Color Table and List View areas in this dialog box shows the details of a particular entity such as its index no, frequency, color value, and so on are given. These details will be displayed on hovering the cursor on each color button.

Note
*This **Palette Manager** tool works only on indexed color and grayscale images. You cannot edit bitonal images using this tool.*

TUTORIALS

Before starting the tutorials you need to download and save the tutorial files on your computer. To do so, follow the steps given next:

1. Download the *c04_rd_2016_tut.zip* file from *http://www.cadcim.com*. The path of the file is as follows:
    Textbooks > Civil/GIS > AutoCAD Raster Design > Exploring AutoCAD Raster Design 2016.

2. Now, save and extract the downloaded folder at the following location:

    C:\AutoCADRasterDesign2016

Notice that *c04_rd_2016_tut* folder is created within the *AutoCADRasterDesign2016* folder.

Tutorial 1   Histogram Tool

In this tutorial, you will perform tonal adjustment on a given image and convert it into a bitonal image using the **Histogram** tool.   **(Expected time: 15 min)**

The following steps are required to complete this tutorial:

a. Start AutoCAD Raster Design and open the new template.
b. Insert the image using the **Insert** tool.
c. Perform tonal adjustment of the raster image using the **Histogram** tool.
d. Convert a non-bitonal image into a bitonal image.
e. Save the image.

Starting AutoCAD Raster Design and Opening the New Template
1. Start AutoCAD Raster Design application and choose **New** from the Application Menu; the **Select template** dialog box is displayed.

2. Select the **acad.dwt** template and choose the **Open** button to open the template file.

Inserting the Image Using the Insert Tool
In this section, you will insert a raster image into the drawing using the **Insert** tool in AutoCAD Raster Design.

1. Choose the **Insert** tool from the **Insert & Write** panel of the **Raster Tools** tab; the **Insert Image** dialog box is displayed.

2. In this dialog box, choose the **Quick insert** radio button from the **Insert Options** area.

3. Next, browse to the location *C:\AutoCADRasterDesign2016\c04_rd_2016_tut\c04_tut01* and then select the **rabi_carto.jpg** file.

4. Next, choose the **Open** button; the **Insert Image** dialog box is closed and the image is inserted into the drawing.

5. Next, enter **Z** at the Command prompt area and press ENTER; you are prompted to specify the zoom option.

6. Enter **E** at the Command prompt area and then press ENTER; the drawing zooms to its extents.

   Note that the image inserted is a multispectral raster image.

Performing Tonal Adjustment of the Raster Image

In this section, you will perform the tonal adjustment of the raster image using the **Histogram** tool.

1. Choose the **Histogram** tool from the **Process Image** drop-down list of the **Edit** panel in the **Raster Tools** tab.

2. Enter **I** at the Command prompt area and press ENTER; the **Histogram** dialog box is displayed.

3. Choose the **Tonal Adjustment** tab from the **Histogram** dialog box, as shown in Figure 4-16.

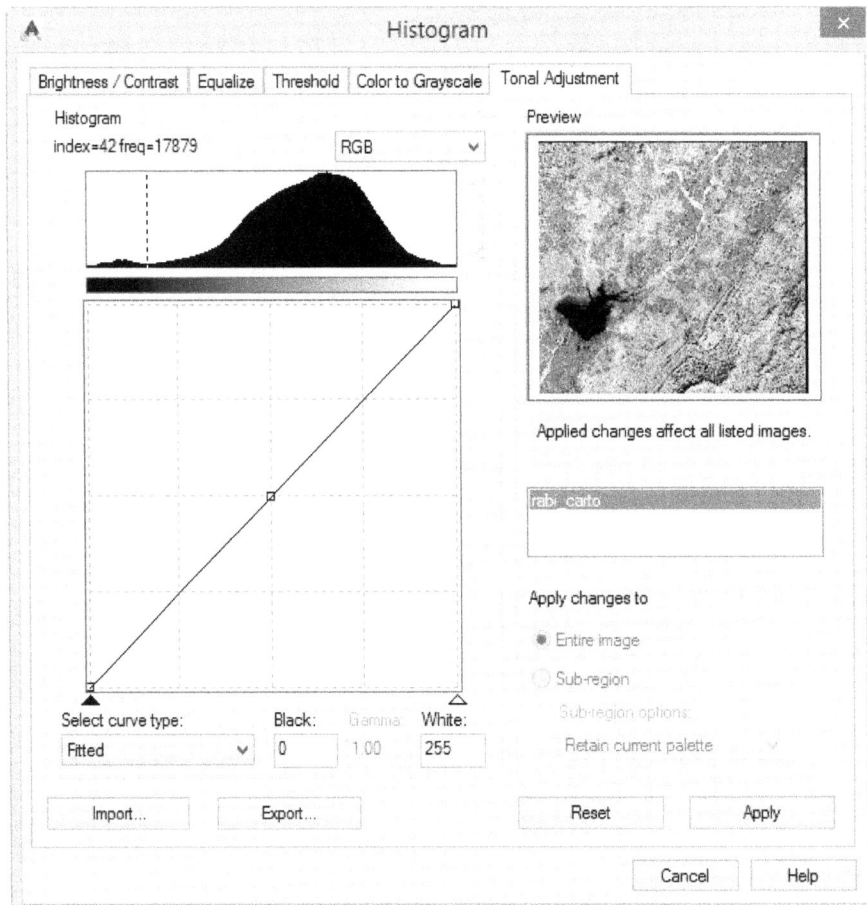

*Figure 4-16 The **Histogram** dialog box with the **Tonal Adjustment** tab chosen*

Note that the histogram of the raster image is not equally distributed.

4. Next, adjust the contrast curve and the other options, as shown in Figure 4-17, and then choose the **Apply** button; the histogram of the image is changed.

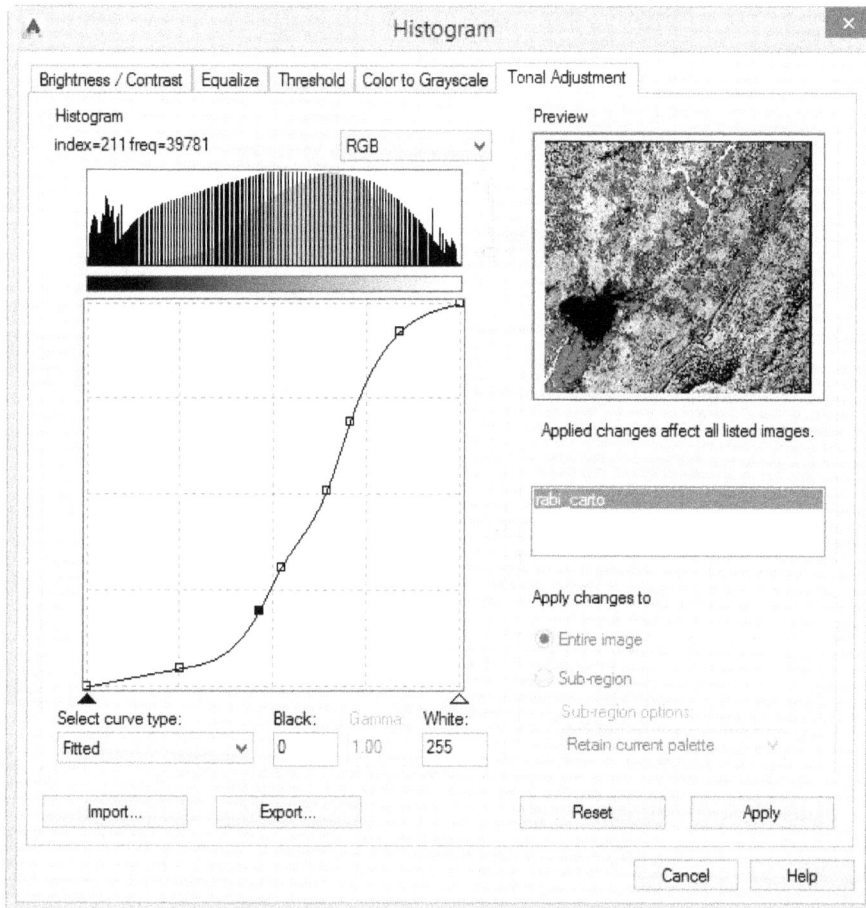

Figure 4-17 The Histogram dialog box with modified contrast curve

5. Choose the **Close** button; the **Histogram** dialog box is closed and the modified image is displayed in the window.

6. Choose the **Save** tool from the **Insert & Write** panel of the **Raster Tools** tab; the image is saved at the specified location.

Converting a Non-Bitonal Image into a Bitonal Image

In this section, you will convert a non-bitonal image into a bitonal image using the **Histogram** tool.

1. Choose the **Histogram** tool from the **Process Image** drop-down list in the **Edit** panel of the **Raster Tools** tab.

2. Enter **I** at the Command prompt area and press ENTER; the **Histogram** dialog box is displayed.

3. Choose the **Threshold** tab from the **Histogram** dialog box, as shown in Figure 4-18.

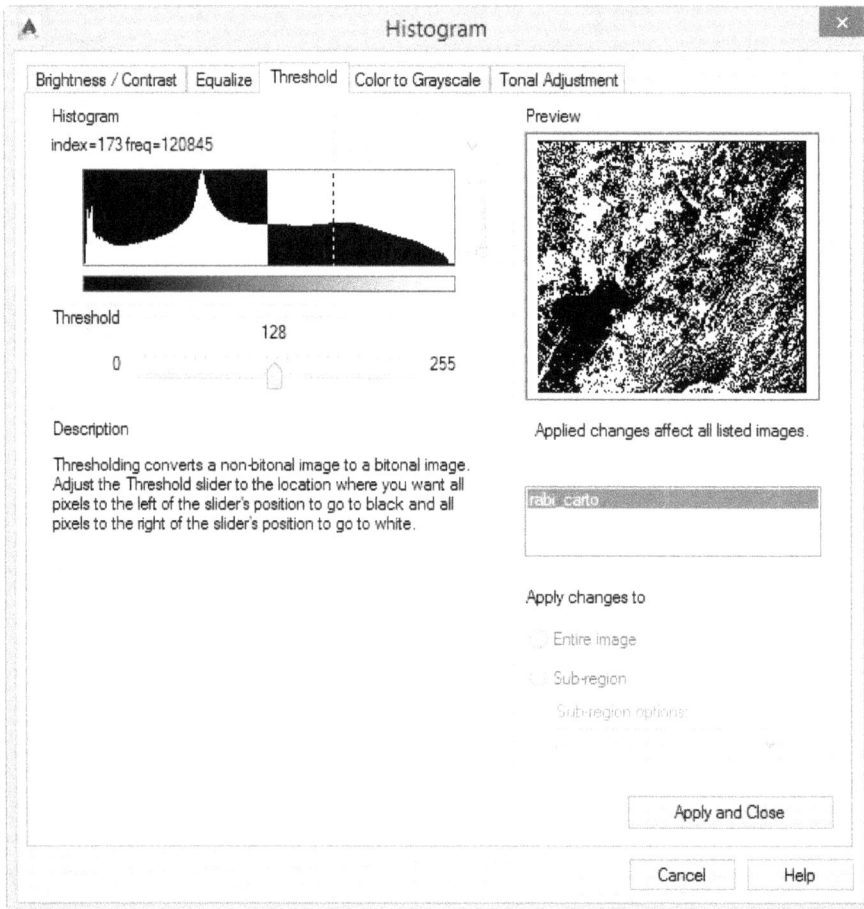

*Figure 4-18 The **Histogram** dialog box with the **Threshold** tab chosen*

4. Choose the **Apply** and then the **Close** button in the dialog box; the **Histogram** dialog box is closed and the image is converted into a bitonal image, as shown in Figure 4-19.

Figure 4-19 *The image after thresholding (bitonal)*

Saving the Image

1. Choose the **Save** tool from the **Insert & Write** panel in the **Raster Tools** tab; the **Save As** dialog box is displayed.

2. In the **Save As** dialog box, browse to the location *C:\AutoCADRasterDesign2016\c04_rd_2016_tut\ c04_tut01* and enter **rabi_carto_bitonal** in the **File name** edit box.

3. Choose the **Save** button; the image file is saved.

4. Now, choose the **Save As** option from the Application Menu; the **Save Drawing As** dialog box is displayed.

5. In this dialog box, enter the file name as **c04_Tut01_Result** and then choose the **Save** button; the image file is saved.

Closing the File

1. Choose the **Close** option from the Application Menu; the file is closed. Ignore the message box displayed, if any.

Tutorial 2   Convolve Tool

In this tutorial, you will apply various filters on the given image using the **Convolve** tool.

(Expected time: 15 min)

The following steps are required to complete this tutorial:

a. Start AutoCAD Raster Design and open the new template.
b. Insert grayscale image using the **Insert** tool.
c. Use median filter to enhance image.
d. Apply highpass filter.
e. Enhance edges in the image.

Starting AutoCAD Raster Design and Opening the New Template

1. Start AutoCAD Raster Design application and choose **New** from the Application Menu; the **Select template** dialog box is displayed.

2. Select the **acad.dwt** template and choose the **Open** button to open the template file.

Inserting the Grayscale Image Using the Insert Tool

In this tutorial, you will insert grayscale image into your drawing for image enhancement.

1. Choose the **Insert** tool from the **Insert & Write** panel of the **Raster Tools** tab; the **Insert Image** dialog box is displayed.

2. In this dialog box, choose the **Quick insert** radio button from the **Insert Options** area.

3. Next, browse to the location *C:\AutoCADRasterDesign2016\c04_rd_2016_tut\c04_tut02* and then select the **rabi_cartogray.jpg** file.

4. Next, choose the **Open** button; the **Insert Image** dialog box is closed and the image is inserted in the drawing.

5. Next, enter **Z** at the Command prompt area and press ENTER; you are prompted to specify the zoom option.

6. Enter **E** at the Command prompt area and then press ENTER; the drawing zooms to its extents.

Enhancing the Image using Median Filter

1. Choose the **Convolve** tool from the **Process Image** drop-down list in the **Edit** panel of the **Raster Tools** tab; the **Image Filters** dialog box is displayed.

2. Expand the **Smoothing Filters** node in the **Select a filter** area and then select the **Median Filter** node, as shown in Figure 4-20.

3. Choose the **Run Filter** button in the **Image Filters** dialog box; the **Median Filter** dialog box is displayed.

4. Enter **5** in the **Filter size (odd integer 3-25)** edit box, refer to Figure 4-21 and then choose the **OK** button; the **Median Filter** dialog box is closed and the image is enhanced, refer to Figure 4-22. You can view the progress bar at the bottom right of the user interface.

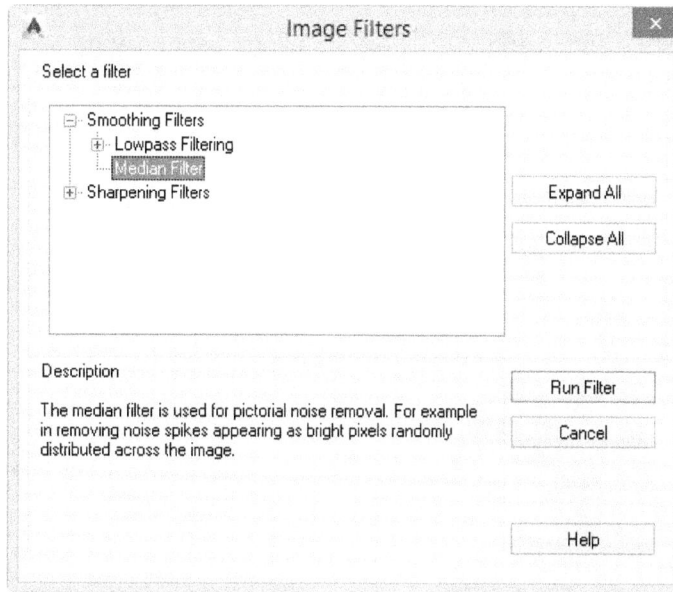

Figure 4-20 The *Image Filters* *dialog box with the* *Median Filter* *chosen*

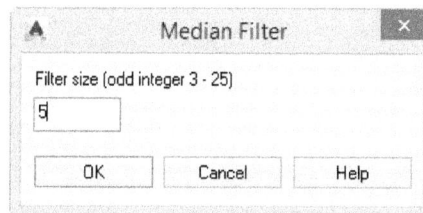

Figure 4-21 The *Median Filter* *dialog box*

Note that the filtering process may take time depending on your system.

5. Choose the **Save** tool from the **Insert & Write** panel of the **Raster Tools** tab; the image is saved in the specified location.

*Figure 4-22 The enhanced image by using the **Median Filter***

Applying Highpass Filters

This process is in continuation of the earlier steps. In this section, you will apply highpass filters to the *rabi_cartogray.jpg* file.

1. Choose the **Convolve** tool from the **Process Image** drop-down list in the **Edit** panel of the **Raster Tools** tab; the **Image Filters** dialog box is displayed.

2. In this dialog box, expand the **Sharpening Filters** node in the **Select a filter** area. Then, choose the **Highpass Filter#1** sub-node under the **Highpass Filters** node, as shown in Figure 4-23.

3. Now, choose the **Run Filter** button in the **Image Filters** dialog box; the dialog box is closed and you can view the progress bar at the bottom right of the user interface.

4. Choose the **Save** tool from the **Insert & Write** panel of the **Raster Tools** tab; the image is saved.

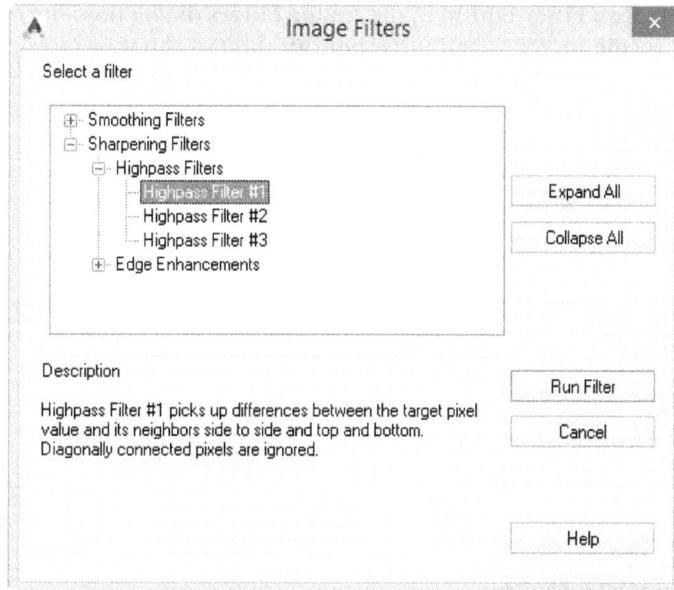

*Figure 4-23 The **Image Filters** dialog box with the **Highpass Filter #1** sub-node selected*

Enhancing Edges in the Image

Now, you will enhance the edges in the given image.

1. Choose the **Convolve** tool from the **Process Image** drop-down list of the **Edit** panel of the **Raster Tools** tab; the **Image Filters** dialog box is displayed.

2. In this dialog box, choose the **Horizontal** sub-node by expanding the **Sharpening Filters > Edge Enhancements > Matched Filter Edge Enhancements** node, as shown in Figure 4-24.

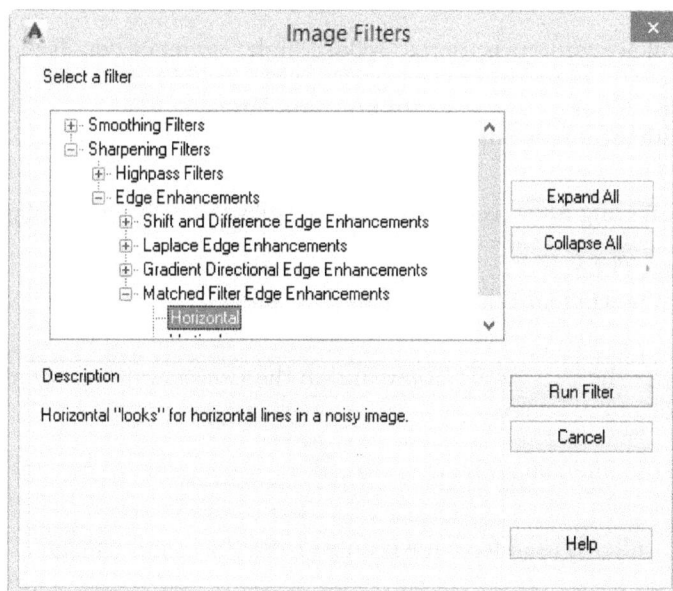

*Figure 4-24 The **Image Filters** dialog box with the **Horizontal** option selected*

3. Now, choose the **Run Filter** button in the **Image Filters** dialog box; the dialog box is closed and you can view the progress bar at the bottom right of the user interface.

4. Choose the **Save** tool from the **Insert & Write** panel of the **Raster Tools** tab; the image is saved at the specified location.

5. Now, choose the **Save As** option from the Application Menu; the **Save Drawing As** dialog box is displayed.

6. In this dialog box, enter the file name as **c04_Tut02_Result** and then choose the **Save** button; the image file is saved.

Closing the File

1. Choose the **Close** option from the Application Menu; the file is closed. Ignore the message box displayed, if any.

Self-Evaluation Test

Answer the following questions and then compare them to those given at the end of this chapter:

1. Which of the following options is used to convert a non-bitonal image into a bitonal image?

   (a) **Threshold**   (b) **Equalize**
   (c) **Tonal Adjustment**   (d) All of the above

2. Which of the following methods is used for image smoothing?

   (a) **Highpass Filter#1**   (b) **Laplace #1**
   (c) **Median Filter**   (d) All of the above

3. Which of the following filters is used to reduce high frequency details in an image?

   (a) **Highpass Filter**   (b) **Lowpass Filter**
   (c) **Edge Enhancement Filter**   (d) All of the above

4. Which of the following filters is used to remove spiky pixels from the image?

   (a) **Highpass Filter**   (b) **Lowpass Filter**
   (c) **Edge Enhancement Filter**   (d) **Median Filter**

5. You need to enter the _____ command in the Command prompt area to change the histogram of the entire image.

6. The process to use filters in an image is called _____.

7. The _____ filter is used to remove spikes from the satellite image.

8. Lowpass filtering removes low frequency values from an image. (T/F)

9. Convolve filters work only on grayscale images. (T/F)

10. Equalizing an image maximizes the details in an image by applying a non-linear contrast stretch technique. (T/F)

Review Questions

Answer the following questions:

1. Thresholding converts a non-bitonal image into a _____ image.

2. The bitonal filters enhance _____ images.

3. The **Indexedcolor** option will be enabled in the _____ area if you insert a bitonal image.

4. _____ can be used to create binary images from the color or grayscale images.

5. To enhance a grayscale image, choose the _____ tool from the **Process Image** drop-down list.

6. Raster pixels store the _____ information of the raster data.

7. You can sharpen an image by using image sharpening filters. (T/F)

8. Median filtering is a linear filtering technique. (T/F)

9. Smoothing an image in AutoCAD Raster Design is a procedure to reduce noise within an image. (T/F)

10. The **Palette Manager** provides information about particular entities such as index number and frequency. (T/F)

11. Highpass filters supress the high frequency values into low frequency values in the targeted pixel. (T/F)

12. Lowpass filters reduce high frequency values of pixels into average values, and results in slightly blurred images. (T/F)

EXERCISES

Exercise 1

Download the *c04_rd_2016_exr.zip* from *http://www.cadcim.com* and convert the given image into grayscale raster image.   **(Expected time: 15 min)**

Exercise 2

Download the *c04_rd_2016_exr.zip* from *http://www.cadcim.com* and equalize the image histogram, refer to Figure 4-25.   **(Expected time: 15 min)**

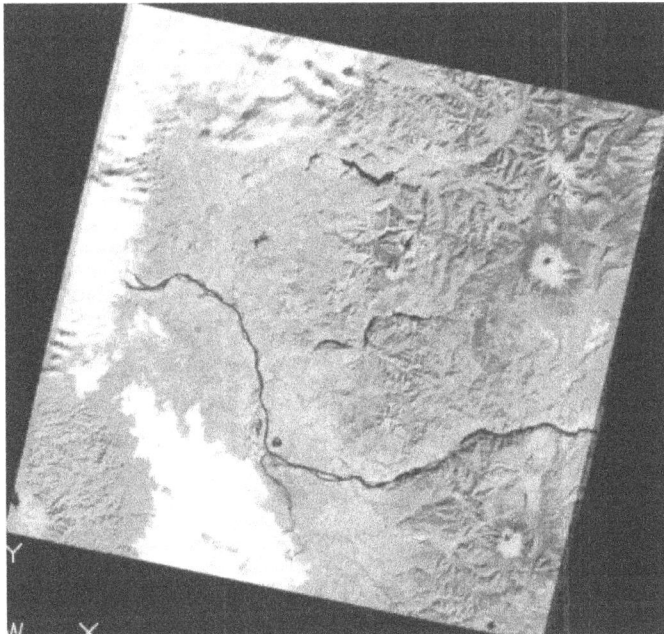

Figure 4-25 *The image to be equalized*

Answers to Self-Evaluation Test

1. a, **2.** c, **3.** b, **4.** d, **5. I, 6.** convolution, **7. Median, 8.** F, **9.** T, **10.** F

Chapter 5

Raster Entity Manipulation (REM) Tools

Learning Objectives
After completing this chapter, you will be able to:
- *Understand Raster Entity Manipulation*
- *Perform basic REM editing*
- *Manipulate REM objects*

INTRODUCTION

This chapter aims to introduce you to the Raster Entity Manipulation (REM) tools of AutoCAD Raster Design. In this chapter, you will learn to edit and manipulate raster images. You will also learn to manipulate raster entities without converting them into vectors.

Moreover, you will learn how to use three types of raster objects to manipulate raster data. This chapter also discusses about the merging of REM objects into raster images as well as editing of bitonal, grayscale, or color images in AutoCAD drawing environment.

UNDERSTANDING RASTER ENTITY MANIPULATION

AutoCAD Raster Design allows you to edit portion of the raster image using the Raster Entity Manipulation (REM) tools. These tools enable you to select lines, arcs, and circles from the raster image and edit those entities. These tools help to edit or manipulate bitonal images from the scanned drawings. Various REM settings and options are discussed next.

Applying REM Settings and Options

Before working with various tools in AutoCAD Raster Design, you need to adjust raster design settings and options. To do so, click on the arrow displayed in the **REM** panel of the **Raster Tools** tab; the **AutoCAD Raster Design Options** dialog box will be displayed, refer to Figure 5-1. Alternatively, you can enter **IOPTIONSPAGE** in the Command prompt area to invoke this dialog box. The tabs and options in the **AutoCAD Raster Design Options** dialog box are discussed next.

REM Tab

In the **AutoCAD Raster Design Options** dialog box, the **REM** tab is selected by default. In this tab, you can specify the settings for the **REM** tools. If you change these settings, the REM objects will change accordingly, as shown in Figure 5-1.

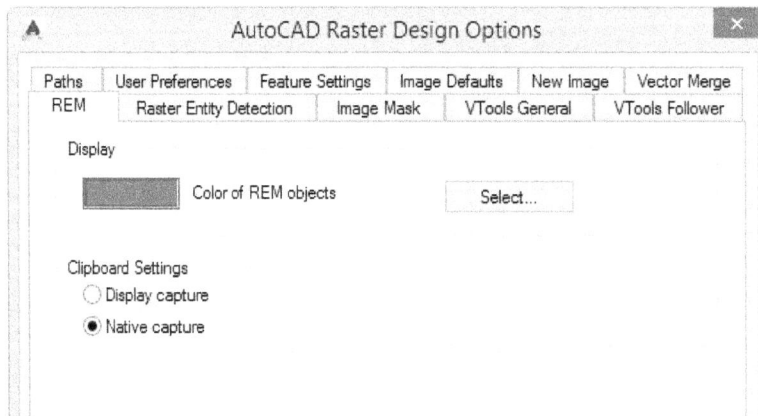

Figure 5-1 *Partial view of the **AutoCAD Raster Design Options** dialog box with **REM** tab chosen*

In the **Display** area, you can specify the color of the REM objects by using the **Color of REM objects** button. Alternatively, you can choose the **Select** button and choose the required color from the palette. In the **Clipboard Settings** area, if you select the **Display capture** radio button, the scale and rotation of the object will not change while copying the raster objects. If you select the **Native capture** radio button, you can change the scale and rotation while copying the raster objects.

Raster Entity Detection Tab

The options in this tab are used to define settings for REM detection tools, refer to Figure 5-2.

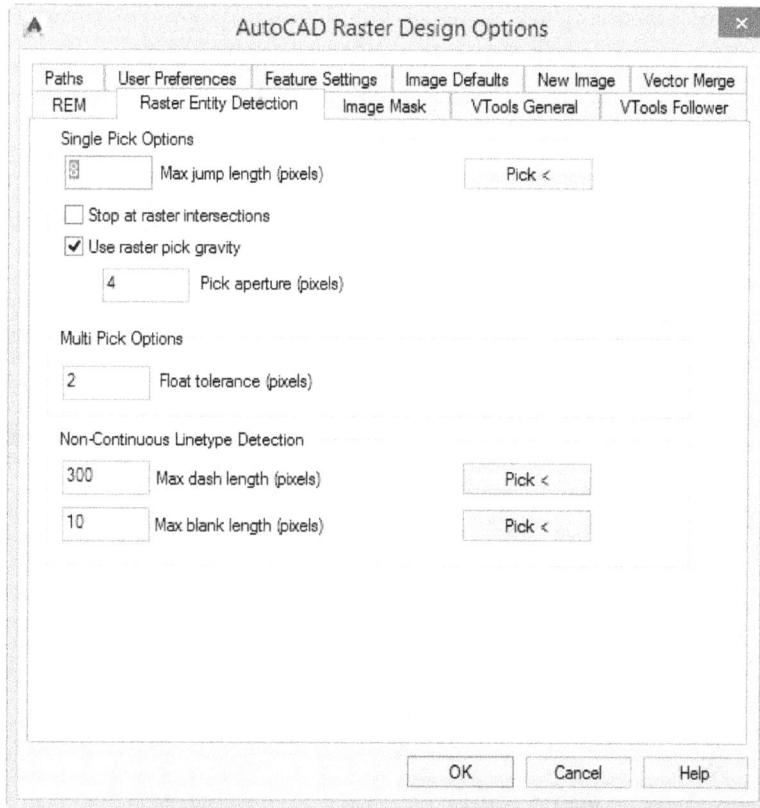

Figure 5-2 The AutoCAD Raster Design Options dialog box with the Raster Entity Detection tab chosen

In the **Single Pick Options** area, specify the distance in the **Max jump length (pixels)** edit box for the raster entity to pick using the **Single Pick** option. If you select the **Stop at raster intersections** check box, the **REM** tools will detect segments or entire primitives during REM operations.

On selecting the **Use raster pick gravity** check box, the **Pick aperture (pixels)** edit box will be enabled. The value entered in the edit boxx will determine the aperture size of the raster entity to be selected.

In the **Multi Pick Options** area, you can specify float tolerance in the **Float tolerance (pixels)** edit box to select raster objects. Similarly, you can specify the parameters for non-continuous linetype detection in the **Non-Continuous Linetype Detection** area.

BASIC REM TOOLS

AutoCAD Raster Design provides set of tools that helps to smoothen an object, cut short raster entity, extend raster entity till specified point, fillet the corner between two entities, merge multiple objects into existing raster. These tools are discussed next.

Knife Tool

Ribbon:	Raster Tools > REM drop-down > Knife
Command:	IKNIFE

You can use the **Knife** tool to separate bitonal raster lines. To separate a bitonal raster object, choose the **Knife** tool from the **REM** drop-down of the **Raster Tools** tab; the cursor will change into a crosshair. Specify the first point inside the border of the image and an end point outside the border of the image, refer to Figure 5-3. Next press ENTER; the raster object will be cut from the entire bitonal image and the command will terminate.

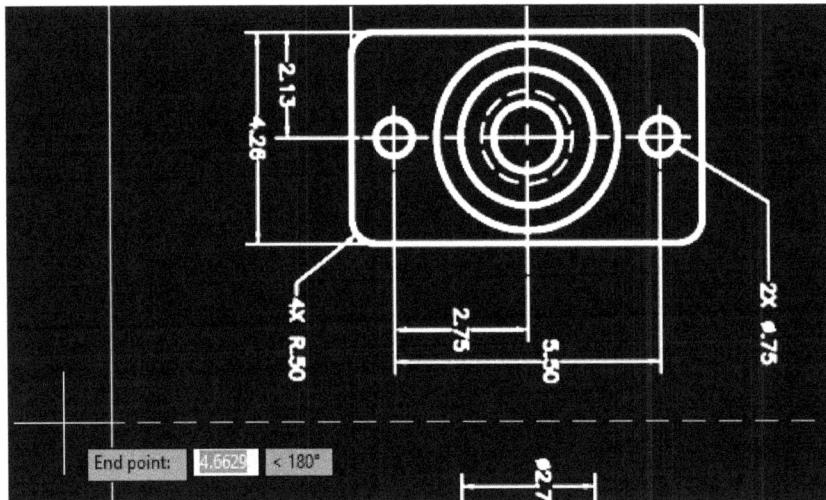

*Figure 5-3 Separating a bitonal image using the **Knife** tool*

Smooth Tool

Ribbon:	Raster Tools > REM drop-down > Smooth
Command:	ISMOOTH

In AutoCAD Raster Design, the **Smooth** tool is used for smoothing a raster feature. This tool converts the discontinuous raster objects into smooth features. To smoothen any raster entity, choose the **Smooth** tool from the **REM** drop-down of the **Raster Tools** tab; the cursor will change into crosshair and you will be prompted to select an object. After selecting the object press ENTER; the REM object will change into a smooth object. Note that sometimes, the **Smooth** tool causes dissociation.

Note

Before smoothing the raster primitive, you must remember to select the desired object as an REM object.

Tip

*You need to use the **Convert To Raster Image** tool after using the **Smooth** tool in order to save changes to the REM objects.*

Trim Tool

Ribbon:	Raster Tools > REM drop-down > Trim
Command:	ITRIM

While working with raster images, you may need to remove unwanted and extended edges. Breaking individual objects takes time if you are working on a complex raster map. In such cases, you can use the **Trim** tool to trim the objects that extend beyond the required point of intersection. To trim the extended object, convert the object by using any of the REM tools. Next, choose the **Trim** tool from the **REM** drop-down of the **Raster Tools** tab; the cursor will change into a selection box and you will be prompted to select cutting edges or boundaries. Next, select the entity that you want to trim; the object will be trimmed, refer to Figure 5-4.

*Figure 5-4 The trim operation done using the **Trim** tool*

Extend Tool

Ribbon:	Raster Tools > REM drop-down > Extend
Command:	IEXTEND

The **Extend** tool may be considered as the opposite of the **Trim** tool. Using this tool, you can extend raster lines and arcs to connect with other objects. However, you cannot extend closed loops. To extend a raster entity, ensure that the desired entity is converted into an REM entity. Next, choose the **Extend** tool; the cursor will change into a selection box. Select the entity to be extended, refer to Figure 5-5; the entity will be extended upto the specified line, refer to Figure 5-6.

Note
*If you want to undo the changes done using the **Extend** tool, enter **U** at the Command prompt area; the changes will be undone and you will be again prompted to select object to extend or shift-select to trim.*

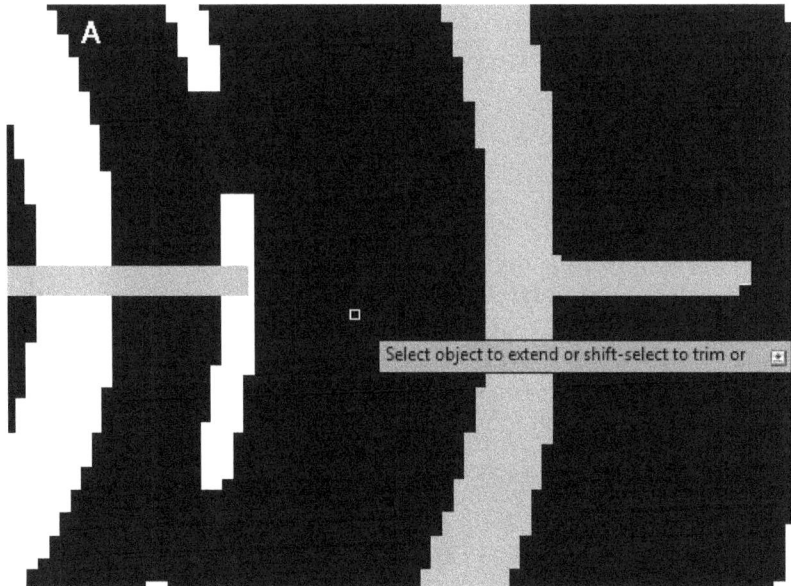

*Figure 5-5 Before performing the **Extend** operation*

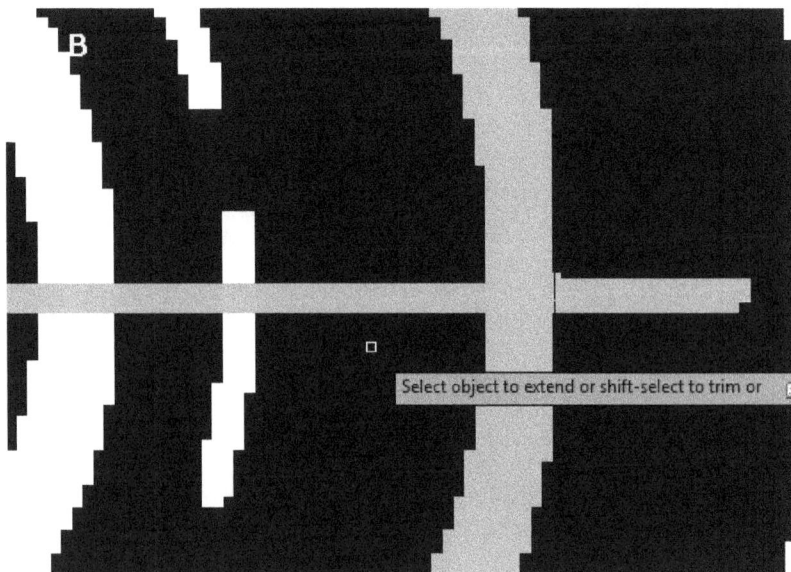

*Figure 5-6 After performing the **Extend** operation*

Fillet Tool

Ribbon: Raster Tools > REM drop-down > Fillet
Command: IFILLET

The **Fillet** tool helps in rounding the corners created between two REM entities. The fillet thus created will be a smooth round shaped arc. A fillet can also be created between two intersecting or REM lines as well as non-intersecting and nonparallel REM lines.

For creating a fillet between two entities, first make sure that the entities are converted into REM entities. Next, choose the **Fillet** tool from the **REM** drop-down of the **Raster Tools** tab; you will be prompted to choose either of the options: **Radius**, **Trim**, or **Multiple**. Choose the **Radius** option from the Command prompt area; you will be prompted to specify fillet radius. Specify the radius and press ENTER; you will be prompted to select the first entity in the drawing window and the **Selection** window will be displayed, as shown in Figure 5-7.

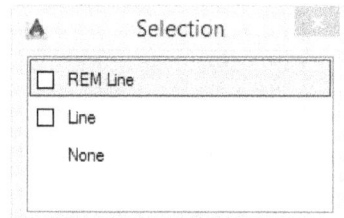

Figure 5-7 *The **Selection** window*

Select the first entity in the drawing window and then select the **REM Line** check box in the **Selection** window and repeat the operation for selecting second entity; the fillet will be created, as shown in Figure 5-8.

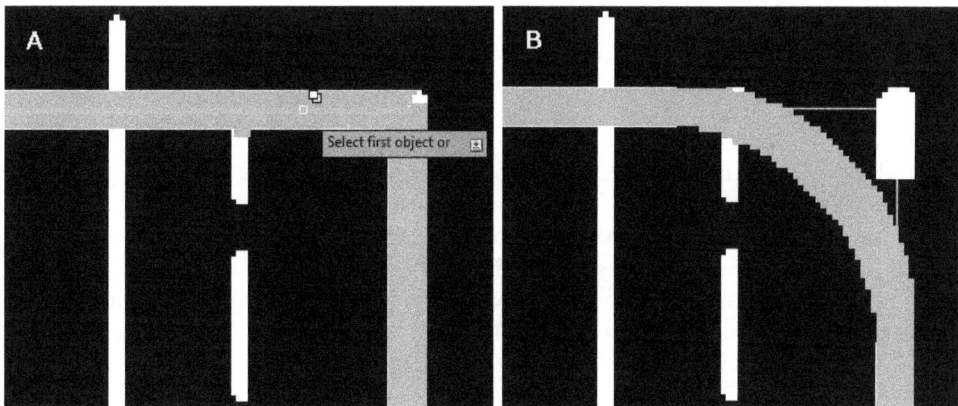

Figure 5-8 *The fillet operation done using the **Fillet** tool*

Offset Tool

Ribbon: Raster Tools > REM drop-down > Offset
Command: IOFFSET

You can use the **Offset** tool to create entities such as lines, polylines, circles, arcs, and curves at a specified offset distance. However, you can offset only one entity at a time. While offsetting a raster object, you need to specify the offset distance and the side to offset. Depending on the side to offset, the resulting object will be smaller or larger than the original object. For example, while offsetting a circle if the offset side is toward the inner side of the perimeter, the resulting circle will be smaller than the original one. To create an entity at specified offset, convert the desired entity into an REM entity and then invoke the **Offset** tool; you will be prompted to specify the offset distance. Specify offset distance in the Command prompt area then press ENTER; you will be prompted to select the object to offset. Select the object; you will be prompted to specify the point on the side to offset. Specify the side to offset; the object will be offset, refer to Figure 5-9.

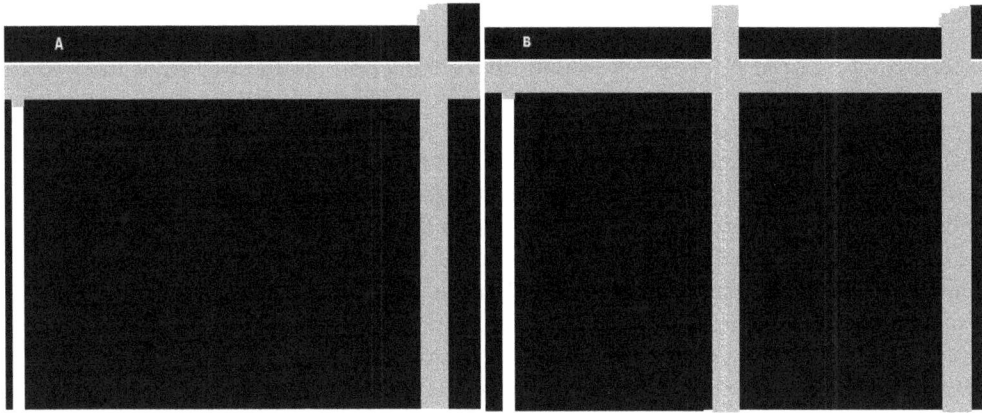

Figure 5-9 *The offset operation done using the **Offset** tool*

Refine Mode Tool

Ribbon:	Raster Tools > REM drop-down > Refine Mode
Command:	IREFINEMODE

In AutoCAD Raster Design, you can modify geometry of raster entities using the **Refine Mode** tool. You can turn on or off the refine mode using the **On/OFf** option for REM objects. To modify raster entities, convert the raster entities into REM primitives and then choose the **Refine Mode** tool from the **REM** drop-down; the cursor will change into a selection box. Next, select the REM objects and press ENTER; you will prompted to select the **On/OFf** option. On choosing the **On** option, the vector entity will be created over raster entity and then you can modify the created vector entity. On choosing the **OFf** option, the entity will remain as an REM object.

Note
After completing the operation you must press ENTER, else the process will continue until it is cancelled.

Remove Data from Region Tool

Ribbon:	Raster Tools > REM drop-down > Remove Data from Region
Command:	IREMOVE

You can use the **Remove Data from Region** tool to delete raster entities from the image. To do so, convert the raster entities into REM entities by choosing any of the REM tools and then choose the **Remove Data from Region** tool, the cursor will change into a selection box and you will be prompted to select the outer boundary of the REM region. Select the outer boundary of the REM region, you will be prompted to **Specify removal method or [?]**. On clicking the **[?]**, the you will be prompted to choose the following options: **SCPolygon/SFence/ SMart/Line/cirCLe/Arc**. Choose the **Line** option; you will be prompted to select one point or 2P method. Select the **2P** option and then specify two points on the line; the selected portion of the REM object will be removed from the raster image, as shown in Figure 5-10.

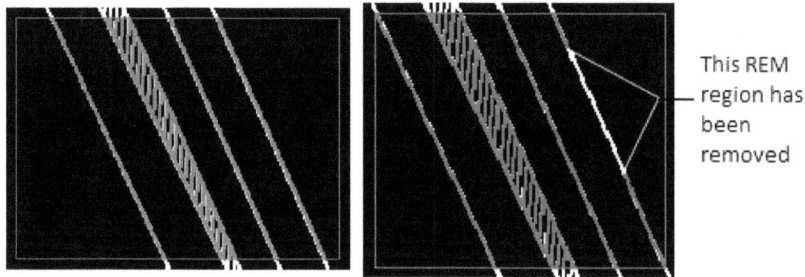

Figure 5-10 *The REM object removed using the 2P method*

Copy to Clipboard Tool

Ribbon: Raster Tools > REM drop-down > Copy to Clipboard
Command: ICOPYSS

Sometimes you need to use same database for various purposes. For this purpose, you need to copy the same object multiple times. In AutoCAD Raster Design, you can copy raster objects to another raster region or clipboard. To do so, ensure that the raster data that you want to copy is selected as an REM object. Then, choose the **Copy to Clipboard** tool from the **REM** drop-down; the cursor will change into a selection box. Select the boundary of the REM object, and then copy and paste the object on the specified location.

REM Background Transparency Tool

Ribbon: Raster Tools > REM panel > REM Background Transparency
Command: ITRANSPARENT

This tool allows you to make the background of the selected objects transparent. REM objects are transparent. Transparency helps you to modify any complex raster drawing. To change the background transparency of any region in a bitonal image, ensure that the raster entities are selected as REM entities within a particular region. Next, choose the **REM Background Transparency** tool from the **REM** panel of the **Raster Tools** tab; the cursor will change into a selection box. Select the boundary and press ENTER or specify Y in the Command prompt area; the background will turn transparent, as shown in Figure 5-11.

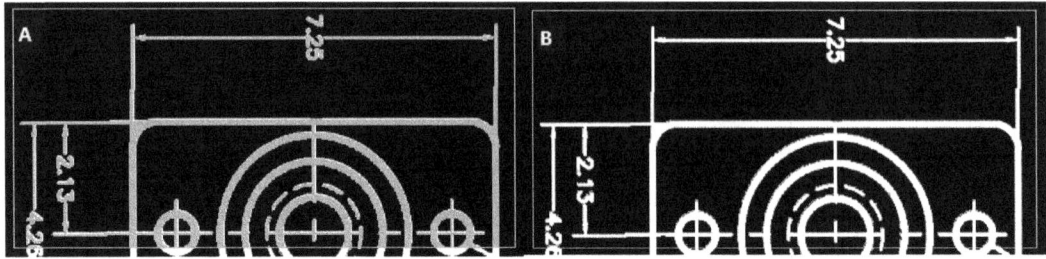

Figure 5-11 *Images before (A) and after (B) using the* ***REM Background Transparency*** *tool*

Merge to Raster Image Tool

Ribbon: Raster Tools > REM panel > Merge to Raster Image
Command: IMERGETOIMG

Sometimes you may need to merge the REM entities to an existing raster image. To do so, ensure that the updated raster objects are selected as REM objects and then choose the **Merge to Raster Image** tool from the **REM** panel of the **Raster Tools** tab; the cursor will change into a selection box. Select the desired REM objects and modify them as per your requirement and then merge them into the original raster image. Next, press ENTER; the selected REM objects will be merged into the raster image.

Tip
If multiple REM objects are opened in different drawing windows, a dialog box will be displayed prompting you to select the image to be modified.

Note
After the completion of the operation, the selected REM objects will be merged into the image and the color of the objects will change into white color.

Select All Tool

Ribbon: Raster Tools > REM panel > Select All
Command: ISELECTALL

The **Select All** tool in AutoCAD Raster Design allows the user to select all REM objects at a time. After selecting the raster objects, you can use a variety of commands to modify the raster entities. To do so, ensure that the desired raster objects are converted as REM objects. Choose the **Select All** tool from the **REM** panel of the **Raster Tools** tab; all REM objects will be selected, as shown in Figure 5-12.

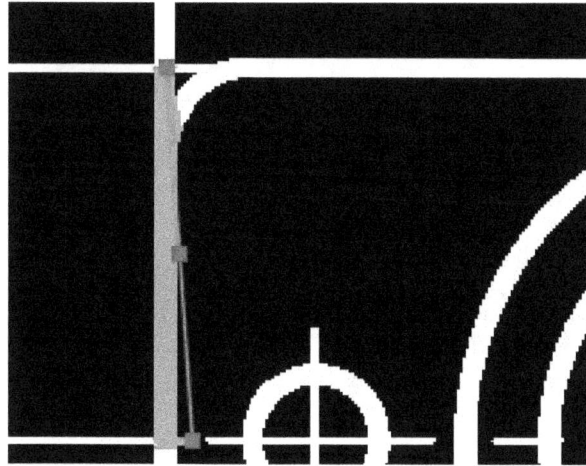

Figure 5-12 *Selecting REM objects using the **Select All** tool*

Clear Selected or Clear All Tool

Ribbon: Raster Tools > REM Panel > Clear Selected or Clear All
Command: ICLEAR or ICLEARALL

In AutoCAD Raster Design, you can clear selected or all REM objects from a raster image by using the **Clear Selected** or **Clear All** tool. The **Clear Selected** tool helps to deselect desired REM objects and the **Clear All** tool helps to deselect all the REM objects from the raster image.

To deselect some desired REM objects, choose the **Clear Selected** tool from the **Clear Selected** drop-down in the **REM** panel of the **Raster Tools** tab; the cursor will change into a selection box. Select the desired REM objects and press ENTER; the selected REM objects will change into normal image elements.

To deselect all REM objects from the image, choose the **Clear All** tool from the **Clear Selected** drop-down in the **REM** panel of the **Raster Tools** tab. On choosing the **Clear All** button, all the selected REM objects will change into raster image elements.

Convert to Raster Image Tool

Ribbon: Raster Tools > REM Panel > Convert to Raster Image
Command: ICONVTOIMG

In AutoCAD Raster Design, you can convert one or more REM objects into a new raster image. This adds up a new raster entity into the drawing. You can copy or edit raster entities from the image and convert them into one or more raster objects. To do so, ensure that the desired image objects are selected as REM objects. Next, choose the **Convert to Raster Image** tool from the **REM** panel of the **Raster Tools** tab; the cursor will change into a selection box and you will be prompted to select an object. Select the REM object from the image and press ENTER; the REM object will convert into a raster object and an outer boundary will be created around the raster object, as shown in Figure 5-13.

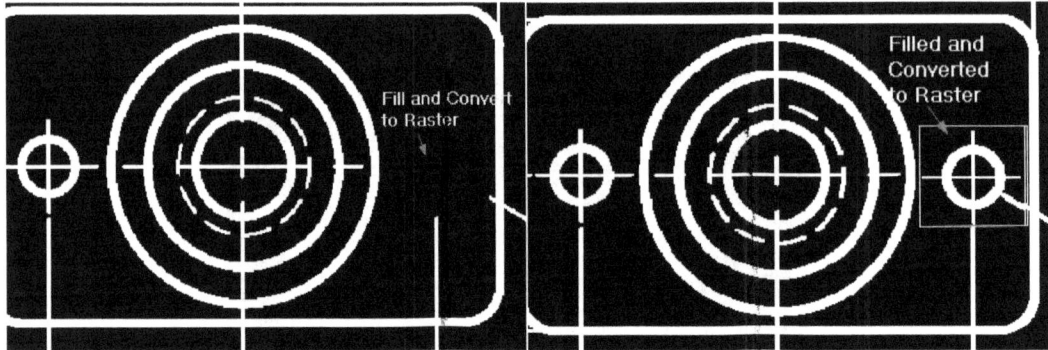

Figure 5-13 *Conversion of REM entities into a new raster image using the **Convert to Raster Image** tool*

> **Tip**
> *If you have selected the raster objects using the **Create Region** or **Enhanced Region** tool, you need to select the outer boundary of the REM objects while converting raster images. And, if you have selected the raster objects using the tools in the **Create Primitive** drop-down, you need to select each object individually.*

MANIPULATION USING REM SELECTION TOOLS

In AutoCAD Raster Design, you can manipulate any raster image for further use by using the tools in the **Create Region**, **Enhanced Region**, and **Create Primitive** drop-downs. Using these tools, you can select raster entities by region or one by one. You can also update raster images without converting them into vectors.

You can manipulate raster images after selecting them as an REM object. You can move, copy, scale, and delete an REM object. Also, you can merge REM objects into an existing image or convert REM objects into a new image. The major functions of the tools available in these drop-downs and the images to be affected by them are given in Tabel 5-1. These tools are discussed in detail in the forthcoming section.

Table 5-1 *REM tools and their functions*

REM Tool	Detail	Image Type
Create Region	Used to select all raster entities within the selected region	Bitonal, grayscale, or color
Enhanced Region	Used to select objects through Smart, Connected, and Connected Entity methods	Bitonal images
Create Primitive	Used to select single raster point, line, or polygon	Bitonal images

Create Region

The **Create Region** drop-down provides various tools to select any raster area or region for any kind of manipulation in the raster data, refer to Figure 5-14. These tools are discussed next.

Polygonal Tool

Ribbon:	Raster Tools > REM Panel > Create Region drop-down > Polygonal
Command:	ISPOLYREG

Sometimes, you need to select raster objects within a polygonal boundary from the raster image for editing or manipulation. You can do so by using the **Polygonal** tool. To select entities within a polygonal boundary, choose the **Polygonal** tool from the **Create Region** drop-down in the **REM** panel of the **Raster Tools** tab; the cursor will change into a crosshair and you will be prompted to specify the first point. Specify the first point. You will be further prompted to specify other points in order to complete the polygonal boundary; keep specifying the other points around the object to complete the polygonal boundary. Next, press ENTER. The desired objects will be selected on the raster image as REM objects, as shown in Figure 5-15.

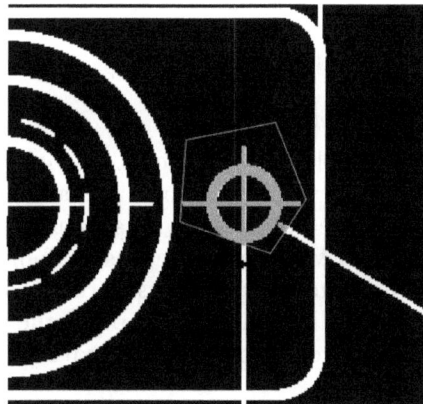

Figure 5-14 The Create Region drop-down

Figure 5-15 Conversion of raster objects to REM entities using the Polygonal tool

Rectangular Tool

Ribbon:	Raster Tools > REM Panel > Create Region drop-down > Rectangular
Command:	ISRECTREG

AutoCAD Raster Design provides tools to select raster objects from the bitonal, color, or grayscale images. The **Rectangular** tool is used to select raster entities from the raster region.

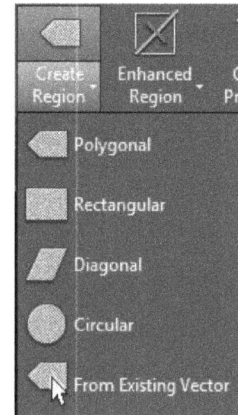

After inserting the image into your drawing, choose the **Rectangular** tool from the **Create Region** drop-down in the **REM** panel of the **Raster Tools** tab. On doing so, the cursor will change into a crosshair and you will be prompted to specify the first corner as well as opposite corner at the desired location on image. Specify the points; a rectangle will be created and the raster objects within the rectangular region will be selected, as shown in Figure 5-16.

Figure 5-16 *Conversion of raster objects to REM entities using the* *Rectangular* *tool*

Diagonal Tool

Ribbon: Raster Tools > REM Panel > Create Region drop-down > Diagonal
Command: ISDIAGREG

The **Diagonal** tool helps to create a diagonal region on a raster image and converts the objects within the diagonal boundary into REM objects. You can modify, delete, and copy these selected objects. To do so, choose the **Diagonal** tool from the **Create Region** drop-down in the **REM** panel of the **Raster Tools** tab; the cursor will change into a crosshair and you will be prompted to specify first corner, adjacent corner, and last corner on the drawing window to define the diagonal region. Specify the points to define the diagonal region; the raster objects will be selected within the diagonal region.

Circular Tool

Ribbon: Raster Tools > REM Panel > Create Region drop-down > Circular
Command: ISCERCREG

The **Circular** tool is used to create a circular region on a raster image and convert the objects within the circular boundary into REM objects. To create a circular region, choose the **Circular** tool from the **Create Region** drop-down in the **REM** panel of the **Raster Tools** tab; the cursor will change into a crosshair and you will be prompted to specify first point and radius of the circle. Specify the first point and the radius to create circle on the desired location; the circle will be created and raster objects within the circular region will be changed into REM objects, as shown in Figure 5-17.

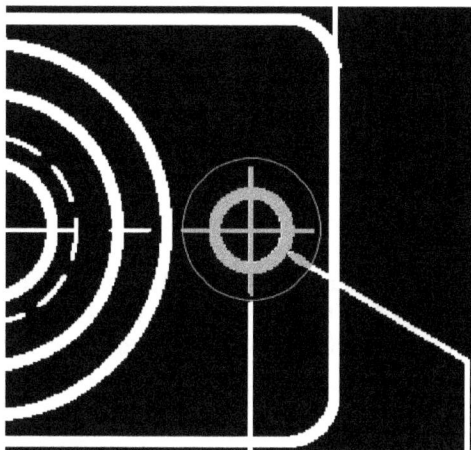

Figure 5-17 *Conversion of raster objects to REM entities using the **Circular** tool*

From Existing Vector Tool

Ribbon: Raster Tools > REM Panel > Create Region drop-down
                   > From Existing Vector
Command: IREGFROMVEC

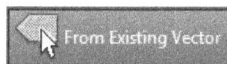

Sometimes, you need to select some raster objects within your area of interest (AOI). This AOI could be the vector boundary of your ward, agriculture field, and so on. You can select raster objects of your AOI by using the **From Existing Vector** tool. To do so, ensure that the raster image and vector boundary of your AOI is open in the drawing. Next, choose the **From Existing Vector** tool from the **Create Region** drop-down in the **REM** panel of the **Raster Tools** tab; the cursor will change into a selection box and you will be prompted to select the closed vector object. Select the closed vector object; the raster objects will be selected within the vector boundary.

Note
Make sure that the vector boundary of the AOI should not be a multiple entity.

Enhanced Region

In AutoCAD Raster Design, you can select the raster objects by defining a enhanced region in the bitonal image. The **Enhanced Region** tools are classified into two categories: Smart and Connected. Smart tools are used to select complete raster entities such as arcs, circles, polygons, and text characters from a raster region. Whereas, connected tools are used to select complete raster entities connected with each other. The **Enhanced Region** tools are discussed next.

Smart Window Tool

Ribbon:	Raster Tools > REM Panel> Enhanced Region drop-down > Smart Window
Command:	ISEBRSMART

The **Smart Window** tool is used to select the raster entities of your desired region by drawing a window around them, as shown in Figure 5-18. As a result, all raster objects that are entirely within the created smart window get selected. To do so, choose the **Smart Window** tool from the **Enhanced Region** drop-down in the **REM** panel of the **Raster Tools** tab; the cursor will change into a crosshair and you will be prompted to specify the first corner point. Specify the first corner point and then specify the opposite corner point in the window; the raster objects will be selected in the window and the objects will be changed into REM objects.

Note
*The **Enhanced Region** tools can be used to select raster objects only on bitonal images.*

Smart Crossing Tool

Ribbon:	Raster Tools > REM Panel > Enhanced Region drop-down > Smart Crossing
Command:	ISEBRSMART

You can use the **Smart Crossing** tool to select raster objects from the bitonal image. This tool helps you to select the raster entities within the smart window.

To select raster entities using this tool, choose it from the **Enhanced region** drop-down in the **REM** panel of the **Raster Tools** tab; the cursor will change into a crosshair and you will be prompted to specify first corner as well as opposite corner in the drawing. On specifying the points, the bitonal objects will change into the REM objects, as shown in Figure 5-19.

*Figure 5-18 Conversion of raster objects to REM entities using the **Smart Window** tool*

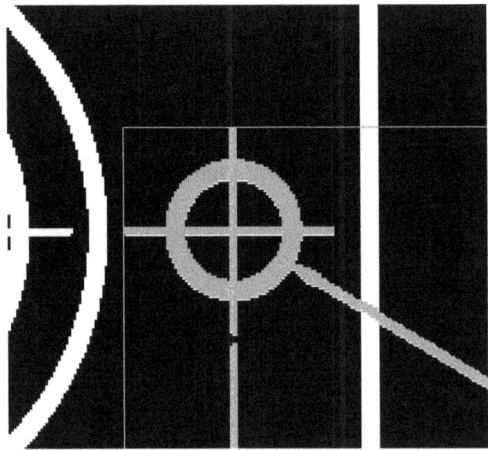

*Figure 5-19 Conversion of raster objects to REM entities using the **Smart Crossing** tool*

Smart Fence Tool

Ribbon:	Raster Tools > REM Panel> Enhanced Region drop-down > Smart Fence
Command:	ISEBRSMART

You can select raster objects by using the **Smart Fence** tool. It creates a smart line around the raster objects and helps to select the desired raster objects.

To create a smart fence for selecting raster entities, choose the **Smart Fence** tool from the **Enhanced Region** drop-down in the **REM** panel of the **Raster Tools** tab; the cursor will change into a crosshair and you will be prompted to specify the first fence point. Specify the first fence point; you will be prompted to specify the endpoint and then press ENTER; the object within the fence will be selected as an REM object.

Note
After selecting the raster objects, a REM boundary will be created around the raster entities.

Connected Entity Tool

Ribbon:	Raster Tools > REM Panel > Enhanced Region drop-down > Connected Entity
Command:	ISEBRCON

Sometimes, you need to select raster objects for manipulation. The **Connected Entity** tool helps you to select raster objects by selecting raster entity. By clicking on a single raster entity, you can select all the connected raster objects.

To select an entity using this tool, choose the **Connected Entity** tool from the **Enhanced Region** drop-down in the **REM** panel of the **Raster Tools** tab. On doing so, the cursor will change into a crosshair and you will be prompted to select a raster entity. Select the raster entity; the selected raster object will change into an REM object. Alternatively, you can select any raster entity on the bitonal image or draw constrain around the object and then select the raster object.

Note
*If the selected object is connected to many objects, you can separate the connected features by using the **Knife** tool.*

Connected Fence Tool

Ribbon:	Raster Tools > REM Panel > Enhanced Region drop-down > Connected Fence
Command:	ISEBRCON

The **Connected Fence** tool helps you to select raster entities by drawing REM lines around the raster entities. This tool is different from the **Connected Entity** tool. It helps in selecting the raster entity by creating a fence across the image. The connected raster entities will change into REM objects in the drawing region of a bitonal image.

To create a connected fence to select raster entities, choose the **Connected Fence** tool from the **Enhance Region** drop-down in the **REM** panel of the **Raster Tools** tab; the cursor will change into a crosshair and you will be prompted to specify first point on the drawing. After specifying the first point, specify the endpoint or line on the image to select raster entities. Next, press ENTER; the desired objects will be selected as REM object, as shown in Figure 5-20.

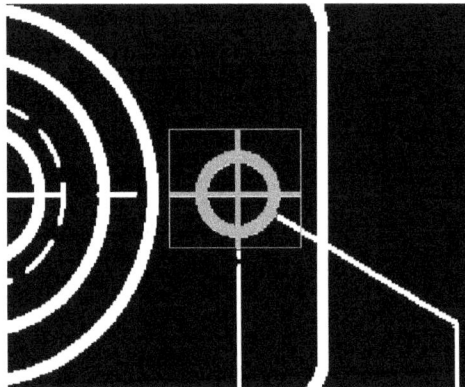

*Figure 5-20 Conversion of raster objects to REM entities using the **Connected Fence** tool*

Create Primitive

AutoCAD Raster Design provides a set of tools in **Create Primitive** drop-down which allow you to distinguish different raster entities such as line, arc, or circle and convert them into vector enties. These tools work only on bitonal images. The funcions of these tools are discussed next .

Smart Pick Tool

Ribbon:	Raster Tools > REM Panel > Create Primitive drop-down > Smart Pick
Command:	ISSMART

Whenever you need to edit or move raster entities, you can select them by using the **Smart Pick** tool. This tool helps to select a raster entity by clicking on the object. Then, you can modify raster entities by using the editing tools.

To edit or move an entity using the **Smart Pick** tool, choose this tool from the **Create Primitive** drop-down in the **REM** panel of the **Raster Tools** tab; the cursor will change into a crosshair and you will be prompted to select a raster entity on the image. Next, click on the desired raster object to select as an REM object, refer to Figure 5-21. You can also select more objects by pressing the ENTER key.

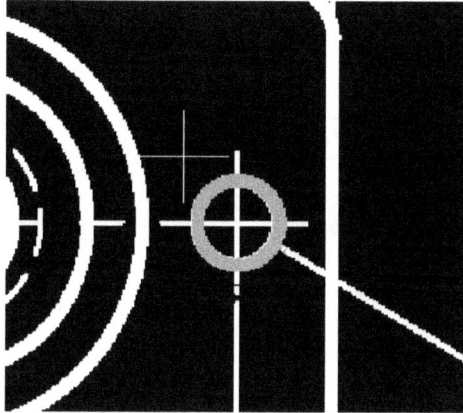

Figure 5-21 Conversion of raster objects to REM entities using the Smart Pick tool

Note
To use the Smart Pick tool, the image must be first converted into a bitonal image.

Line Tool

Ribbon:	Raster Tools > REM Panel > Create Primitive drop-down > Line
Command:	ISLINE

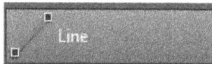

Sometimes, you need to edit or move raster entities from an image. The **Line** tool helps to select line feature from the image and change them into REM primitive. Then, you can edit or manipulate raster objects on the image. To do so, choose the **Line** tool from the **Create Primitive** drop-down in the **REM** panel of the **Raster Tools** tab; the cursor will change into a crosshair. Select any line on the image; the raster line will change into REM primitive, as shown in Figure 5-22. Now, you can edit or modify the selected REM primitives. You can select multiple raster entities by pressing the ENTER key.

Note
After choosing the Line tool, if you select a geometric object other than line, a message will be displayed informing that the selected entity is not a line.

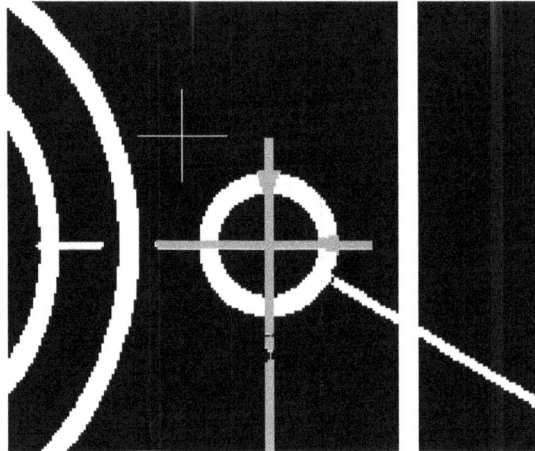

*Figure 5-22 Conversion of raster objects into REM entities using the **Line** tool*

Circle Tool

Ribbon: Raster Tools > REM Panel > Create Primitive drop-down > Line
Command: ISCIRCLE

The **Circle** tool is used to edit or modify only circle object. Using this tool, you can change the raster object into REM primitive and then edit them according to your needs. To do so, choose the **Circle** tool from the **Create Primitive** drop-down in the **REM** panel of the **Raster Tools** tab; the cursor will change into a crosshair. Select any circular object on the image; the raster circle will change into an REM primitive, as shown in Figure 5-23. Next, you can copy or edit these REM primitives.

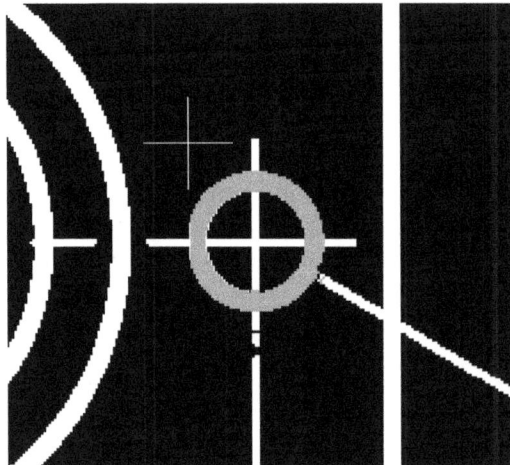

*Figure 5-23 Conversion of raster objects into REM entities using the **Circle** tool*

For example, if you select any line entity after choosing the **Circle** tool, a message will be displayed informing that the selected entity is a line entity not a circle.

Note
You can select circle entities by using the one-pick point, Centre, 2P, and 3P methods.

Arc Tool

Ribbon:	Raster Tools > REM Panel > Create Primitive drop-down > Arc
Command:	ISARC

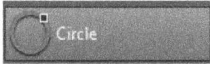

The **Arc** tool helps to select arc objects from the bitonal image. To do so, choose the **Arc** tool from the **Create Primitive** drop-down in the **REM** panel of the **Raster Tools** tab; the cursor will change into a crosshair. Select the desired arc object to be selected as an REM primitive object; the raster entity will change into REM primitive and will be ready for modification.

Note
You can select arc entities by using the one-pick point, Centre, and 3P methods.

TUTORIALS

Before starting the tutorial, you need to download and save the tutorial files on your computer. To do so, follow the steps given next.

1. Download the *c05_rd_2016_tut.zip* file from *http://www.cadcim.com*. The path of the file is as follows:
    Textbooks > Civil/GIS > AutoCAD Raster Design > Exploring AutoCAD Raster Design 2016.

2. Now, save and extract the downloaded folder at the following location:

    C:\AutoCADRasterDesign2016

Notice that *c05_rd_2016_tut* folder is created within the *AutoCADRasterDesign2016* folder.

Tutorial 1   Using REM Tools

In this tutorial, you will modify a raster image using REM commands.

(Expected time: 30 min)

The following steps are required to complete this tutorial:

a. Start AutoCAD Raster Design application and open the drawing file.
b. Copy raster region using REM commands.
c. Delete raster regions.
d. Save the current image.

Starting AutoCAD Raster Design and Opening the Drawing File

1. Start AutoCAD raster Design application and choose the **Open** button from the Quick Access Toolbar; the **Select File** dialog box is displayed.

2. In this dialog box, browse the following location:

    C:\AutoCAD Raster Design 2016\c05_rd_2016_tut\c05_tut01

3. Select the **c05_Tut01.dwg** file; the preview of the selected drawing file is displayed in the **Preview** area and *c05_Tut01.dwg* is displayed in the **File name** edit box.

4. Choose the **Open** button; the **Select File** dialog box is closed and the drawing is displayed in the drawing window.

Copying Raster Entities Using the Create Region Tools

1. Zoom in on the raster image so that the raster image is clearly visible in the drawing window, as shown in Figure 5-24.

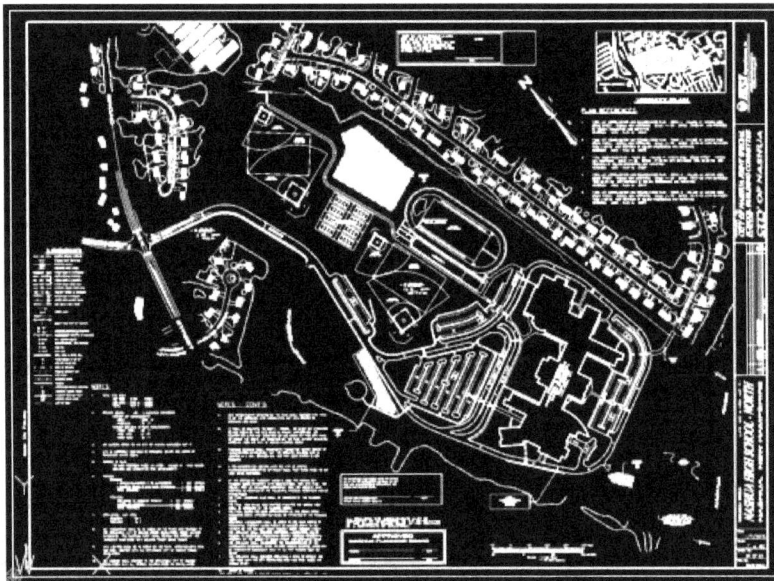

Figure 5-24 Raster image to be modified

2. Choose the **Rectangular** tool from the **Create Region** drop-down in the **REM** panel of the **Raster Tools** tab; the cursor is changed into a crosshair and you are prompted to specify first corner point in the drawing.

3. Click at the upper-left corner of the drawing, as shown in Figure 5-25; you are prompted to specify the other corner point.

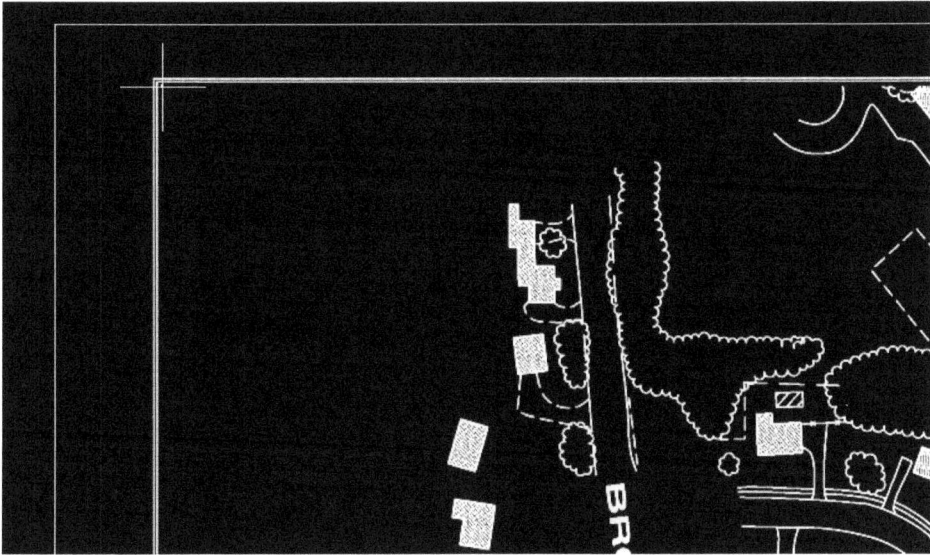

Figure 5-25 Upper-left corner on the drawing specified

4. Click in the lower-right corner of the drawing, refer to Figure 5-26.

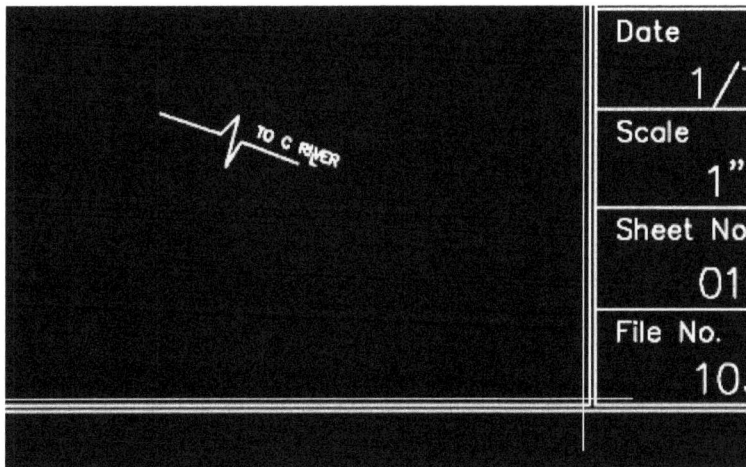

Figure 5-26 Lower-right corner of the drawing specified

Note that the raster region is now selected as REM entities and the color of the raster entities is changed.

5. Choose the **Select All** button from the **REM** panel of the **Raster Tools** tab.

6. Ensure that the boundary of the REM region is selected. Enter **COPY** in the Command prompt area and then press ENTER; you are prompted to specify the base point.

7. Click at the lower left corner of the REM drawing boundary; you are prompted to specify the second point.

8. Click at the lower-right corner of the REM drawing boundary and press ENTER, refer to Figure 5-27.

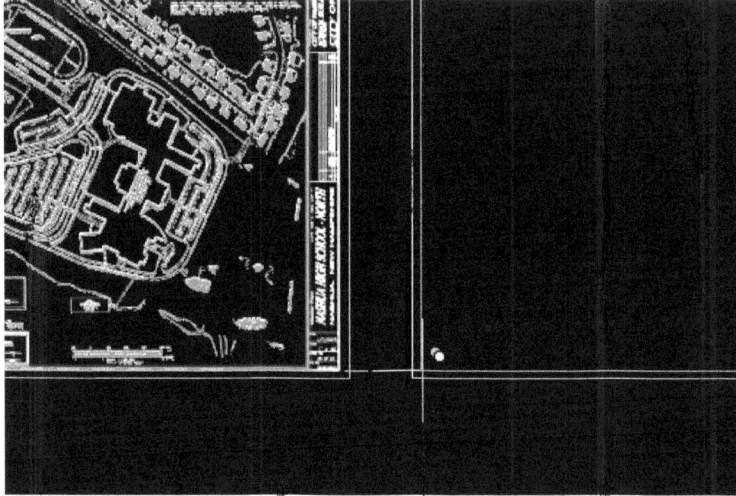

Figure 5-27 Lower-left corner of the drawing rectangle

Note that the REM entities are copied to the new area in the drawing window.

9. Choose the **Convert to Raster Image** button from the **REM** panel of the **Raster Tools** tab; the cursor is changed into a selection box.

10. Select the newly created REM boundary and press ENTER; the REM entities are now converted into raster image, as shown in Figure 5-28. Note that the REM entities are converted into raster image and the color of the raster objects is changed into default color (white).

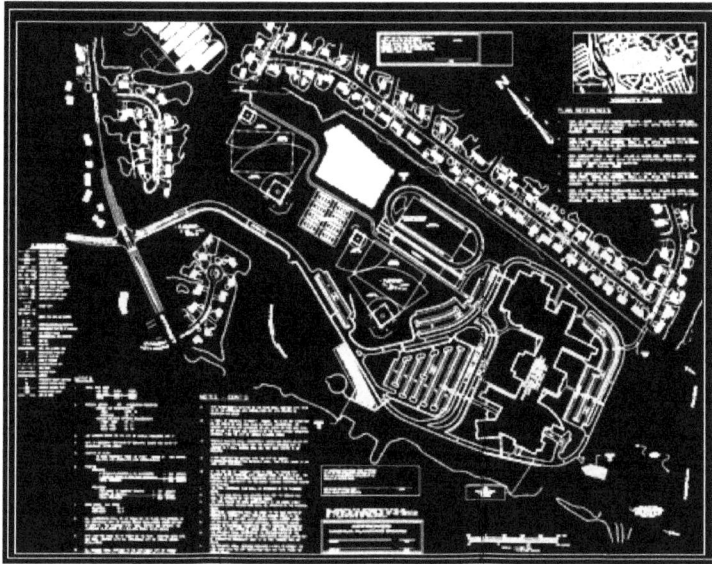

Figure 5-28 Raster image copied to a new location

Deleting Raster Region from the Image

1. Enter **ERASE** in the Command prompt area and press ENTER; the cursor is changed into a selection box.

2. Select the outer boundary of the previous image and press ENTER; the image is deleted.

3. Next, zoom in on the specified location in the newly created REM entities, as shown in Figure 5-29.

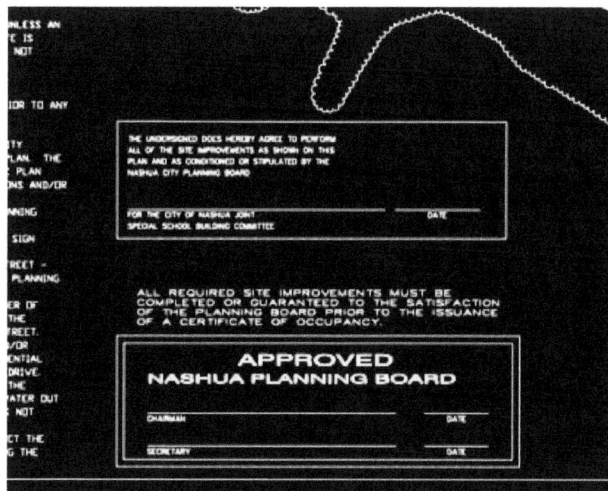

Figure 5-29 Zoomed region of the image

4. Select the **Polygonal** tool from the **Create Region** drop-down in the **REM** panel; the cursor is changed into a crosshair.

5. Specify the points in the drawing area such that the raster region is enclosed in the polygon, refer to Figure 5-30, and press ENTER; the raster region is selected as REM region and the boundary is created.

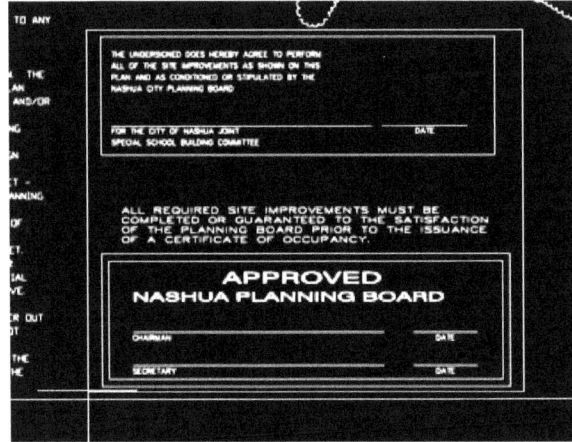

Figure 5-30 Specified REM region

6. Select the newly created REM boundary and press DELETE; the REM region is deleted.

Saving the File

1. Choose **Save As** from the Application Menu; the **Save Drawing As** dialog box is displayed. Ignore if any message box is displayed.

2. In this dialog box, browse to the following location:

    C:\AutoCADRasterDesign2016\c05_rd_2016_tut\c05_tut01

3. In the **File name** edit box, enter **c05_Tut01_Result**.

4. Choose the **Save** button; the dialog box is closed and the drawing file is saved with the name **c05_Tut01_Result.dwg** at the specified location. Also, the **Save Image** dialog box is displayed, as shown in Figure 5-31, informing you that the raster in the drawing has been modified.

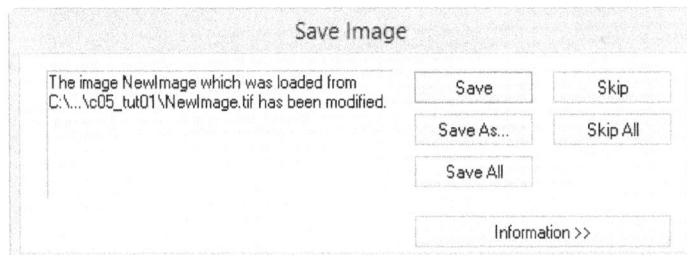

*Figure 5-31 The **Save Image** dialog box*

5. Choose the **Save As** button from the dialog box; the **Save As** dialog box is displayed.

6. In this dialog box, browse to the following location:

    C:\AutoCADRasterDesign2016\05_rd_2016_tut\c05_tut01

7. In the **File name** edit box, enter **Master_Site_Plan_Correlated.**

8. Choose the **Save** button; the **Save As** dialog box is closed and the **Encoding Method** dialog box is displayed, as shown in Figure 5-32.

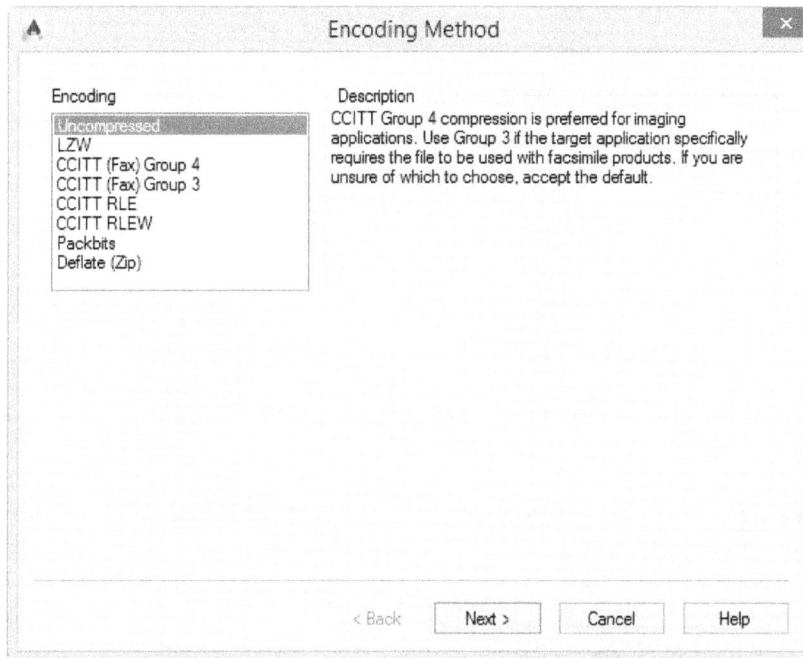

*Figure 5-32 The **Encoding Method** dialog box*

9. Choose the **Uncompressed** option from the **Encoding** list in this dialog box and then choose the **Next** button; the **Data Organization** dialog box is displayed, as shown in Figure 5-33.

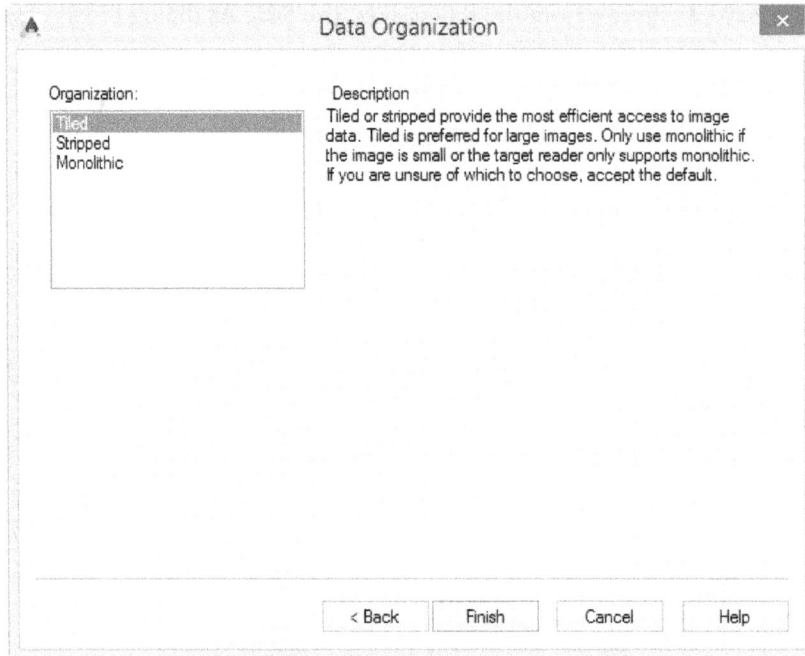

Figure 5-33   *The* **Data Organization** *dialog box*

10. Choose the **Tiled** option from the **Organization list** dialog box and then choose the **Finish** button; the data is exported to the specified location.

Closing the File

1. Choose the **Close** option from the Application Menu; the file is closed. Ignore the message box displayed, if any.

Tutorial 2   Using REM Operations Tools

In this tutorial, you will correct imperfections in bitonal drawings.

(Expected time: 30 min)

The following steps are required to complete this tutorial:

a. Open AutoCAD Raster Design application and insert the image using the **Insert** tool.
b. Extend the raster entities and modify the drawing.

Starting AutoCAD Raster Design and Opening the New Template

1. Start AutoCAD Raster Design application and choose **New** from the Application Menu; the **Select Template** dialog box is displayed.

2. Select the **acad.dwt** template and choose the **Open** button to open the template file.

Inserting the Image Using the Insert Tool

1. Choose the **Insert** tool from the **Insert & Write** panel of the **Raster Tools** tab; the **Insert Image** dialog box is displayed.

2. In this dialog box, choose the **Quick insert** radio button in the **Insert Options** area.

3. Browse to the location *C:\AutoCADRasterDesign2016\c05_rd_2016_tut\c05_tut02* and select the **Master Site Plan-Master Plan.tif** file.

4. Choose the **Open** button; the **Insert Image** dialog box is closed and the image is inserted in the drawing.

5. Specify **Z** in the Command prompt area and press ENTER; you are prompted to specify the zoom option.

6. Specify **E** and then press ENTER; the drawing is zoomed to its extents.

Extending the Line Feature

1. Zoom to the specified location, as shown in Figure 5-34.

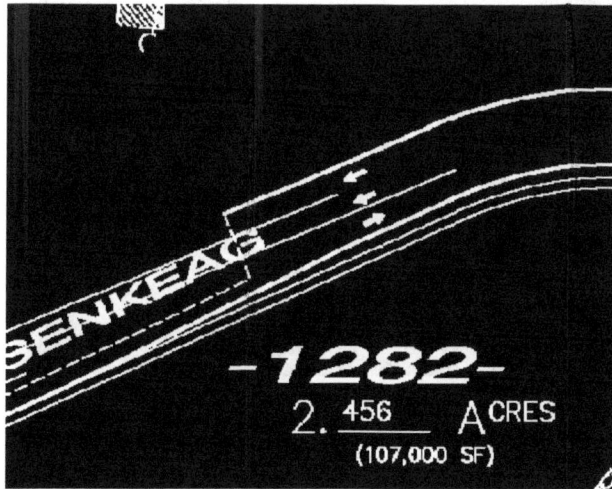

Figure 5-34 Image area to be zoomed

2. Choose the **Smart Pick** tool from the **Create Primitive** drop-down in the **REM** panel of the **Raster Tools** tab; the cursor is changed into a crosshair.

3. Select the line to be extended; an REM line is created over the raster line, refer to Figure 5-35.

4. Right-click and choose **Repeat ISSMART**.

5. Select the boundary edge for the extended line, as shown in Figure 5-36.

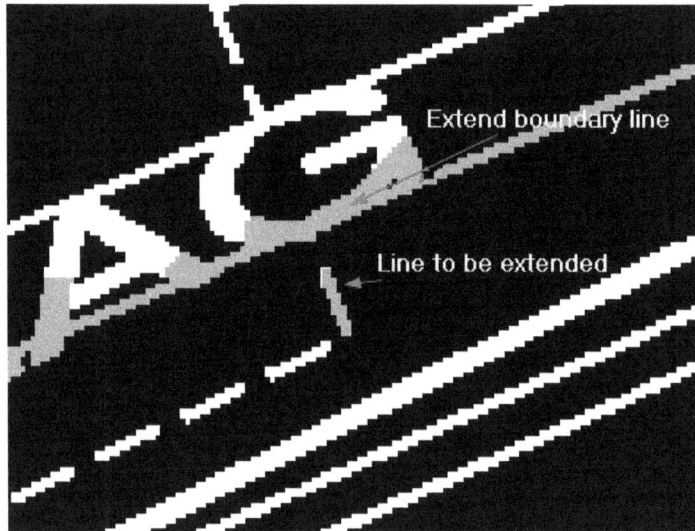

Figure 5-35 Raster line to be extended

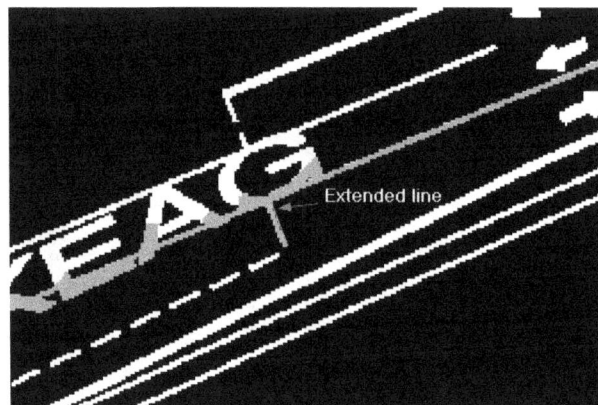

Figure 5-36 The extended line

6. Choose the **Extend** tool from the **REM** drop-down in the **REM** panel of the **Raster Tools** tab; the cursor is changed into a selection box.

7. Select the boundary edge for the extended line and press ENTER.

8. Select the line to be extended and press ENTER to end the command.

9. Choose the **Convert to Raster Image** tool and select the REM entities. Press ENTER; the REM objects get merged with the raster image.

10. Choose the **Save** button from the **Insert & Write** panel; the image is saved.

Closing the File

1. Choose the **Close** option from the Application Menu; the file is closed. Ignore the message box displayed, if any.

Self-Evaluation Test

Answer the following questions and then compare them to those given at the end of this chapter:

1. Which of the following options is used to shorten REM entities from a raster image?

    (a) **Extend**   (b) **Trim**
    (c) **Offset**   (d) All of the above

2. Which of the following tools copies REM objects to the clipboard?

    (a) **Select All**   (b) **Copy to Clipboard**
    (c) **Refine Mode**   (d) None of these

3. Which of the following tools is used to select a rectangular region?

    (a) **Smart Pick**   (b) **Polygonal**
    (c) **Rectangular**   (d) All of the above

4. Which of the following tools is used to create REM enhanced bitonal region by fence selection?

    (a) **Connected Fence**   (b) **Connected Window**
    (c) **Smart Window**   (d) **Smart Fence**

5. You can extend a raster entity using the _____ tool.

6. REM objects can be merged into existing images. (T/F)

7. AutoCAD Raster Design allows you to create three types of REM objects. (T/F)

8. The REM tools work on grayscale or color images. (T/F)

Review Questions

Answer the following questions:

1. The **Connected Entity** tool is used for selecting _____ entities connected to a specified point.

2. You can define enhanced bitonal region objects only in _____ images.

3. The _____ tool is used to select all the raster entities that are connected to entities within the window.

4. The _____ tool is used to create a REM enhanced region in a bitonal image.

5. You can update raster images without converting them into vectors. (T/F)

6. You can define region objects in bitonal, grayscale or color images. (T/F)

7. You can not change the color of REM objects. (T/F)

8. After selecting raster objects using region objects, raster images will change into vector entities. (T/F)

EXERCISE

Exercise 1

Download the *c05_rd_2016_exr.zip* from *http://www.cadcim.com* and shorten the raster lines, for the given image as shown in Figure 5-37.   **(Expected time: 30 min)**

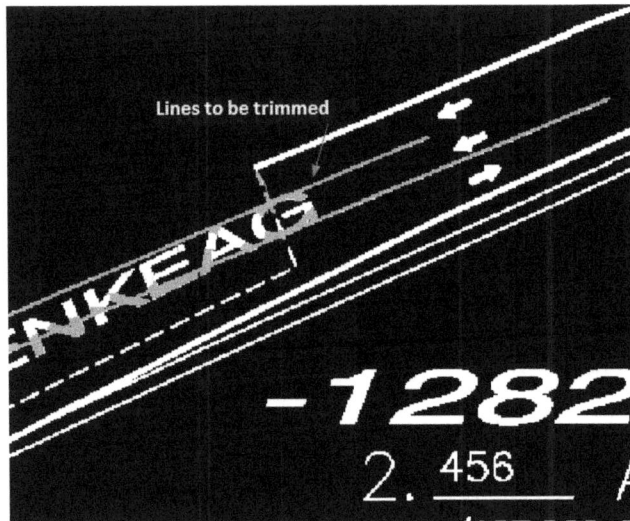

Figure 5-37 Lines to be trimmed

Chapter 6

Vectorization Tools

Learning Objectives

After completing this chapter, you will be able to:

• *Understand raster to vector conversion concept*
• *Use the Followers tools*
• *Work with the Primitives tools*
• *Use the Text Manipulation tools*
• *Apply the OCR tools*

INTRODUCTION

Raster and vector are two basic data structures that are used for storing and manipulating spatial information. Raster data is represented in the form of pixels whereas vector data is represented in the form of points, lines, and polygons that are geometrically and mathematically associated.

This chapter discusses about raster to vector conversion tools given in AutoCAD Raster Design. Also, you will learn about editing and adjustment of raster images and manipulating raster entities by converting them into vector objects.

Moreover, you will learn to convert raster polylines into 2D vector polylines, create contours from scanned topo maps, replace raster text, delete older text from raster image, and edit multiple raster texts.

RASTER TO VECTOR CONVERSION

Database from raster images are generally created using techniques like digitization to draw information from the images. These techniques are very tedious and time consuming. AutoCAD Raster Design provides interactive tools to convert raster entities into vector entities easily and quickly.

Vector data is a mathematical description of geometry. Vector entities can be a point, line, or polygon feature. Raster-to-vector conversion of a drawing archive is a resource-intensive undertaking that often requires large amount of time and money. To maintain the usability of the drawings, the drawing need to be converted into CAD drawing format. Conversion can be carried out on the gray scale scan using image enhancement and image processing tools that are necessary for clarifying raster details.

AutoCAD Raster Design provides vectorization tools like the **Primitives**, **Followers**, **Text**, and **OCR** tools. The **Primitives** and **Followers** tools are used to convert raster geometry into vector geometry in the drawings whereas the **Text** and **OCR** tools are exclusively used to convert any text in the given drawing into vector text. Before using these vectorization tools, you need to define general settings in the **VTools General** and **VTools Follower** tabs in the **AutoCAD Raster Design Options** dialog box. Major options in the **VTools General** and **VTools Follower** tabs have been discussed next.

VTools General Tab

To define the settings in the **VTools General** tab, choose the **Options- VTools General** button from the **Vectorize & Recognize Text** panel of the **Raster Tools** tab; the **AutoCAD Raster Design Options** dialog box will be displayed with the **VTools General** tab chosen.

In this tab, you can will define the settings for the vectorization tools which will inturn define the vector objcets to be created. Various options in this tab are discussed next, refer to Figure 6-1.

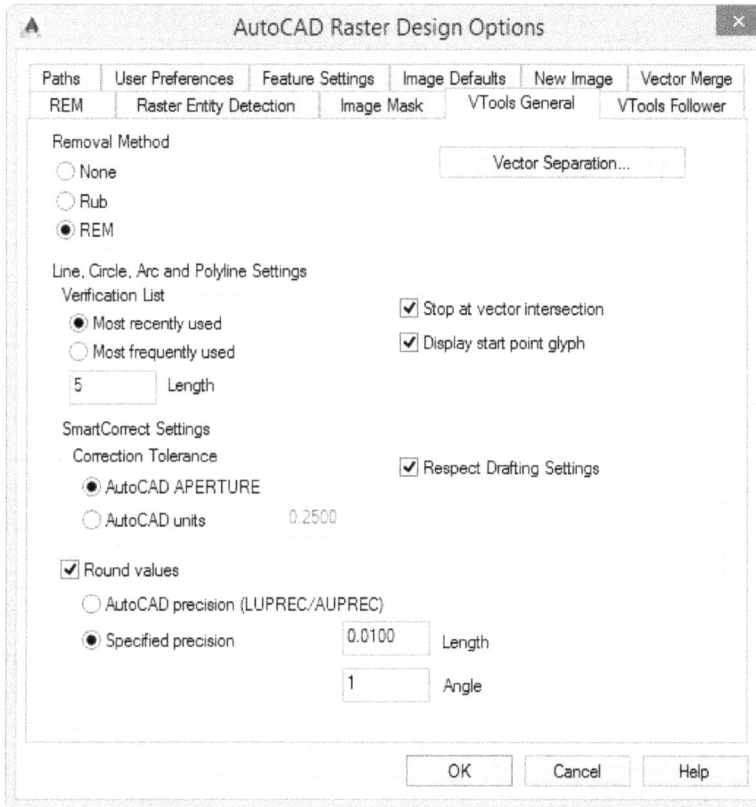

*Figure 6-1 The **AutoCAD Raster Design Options** dialog box with the **VTools General** tab chosen*

In the **Removal Method** area of the **VTools General** tab, if you select the **None** radio button, the underlying raster will remain intact. On selecting the **Rub** radio button, the underlying raster entities will automatically get deleted. On selecting the **REM** radio button, the underlying raster entities get deleted but raster intersections remain intact.

The **Vector Separation** button helps specifying values for layer or polyline width based on the underlying raster. If you choose this button, the **Vector Separation Options** dialog box will be displayed. In this dialog box, choose the **Contour** tab to specify vector line widths for contours based on their elevation interval. You can also choose the **General** tab for specifying vector line widths, layer assignments, and the handling of non-continuous lines. Choose the **OK** button to close the **Vector Separation Options** dialog box.

The **Line, Circle, Arc and Polyline Settings** area is further divided into two areas: **Verification List** and **SmartCorrect Settings**. The **Line, Circle, Arc and Polyline Settings** area allows you to set the values for line, circle, and arc settings.

In the **Verification List** area you can verify the **Most recently used** and the **Most frequently used** tools. If line, arc, circle, and polyline are used recently, then choose the **Most recently used** radio button, else, you can choose the **Most frequently** used radio button. Further, you can also verify the values of the length in the **Length** edit box.

The **SmartCorrect Settings** area is used to specify the correlation tolerance and then the round values for the vector conversion. You can also set values of length and angle in the corresponding edit boxes.

VTools Follower Tab

Using the **VTools Follower** tab, you can set the behavior of the follower tools, refer to Figure 6-2.

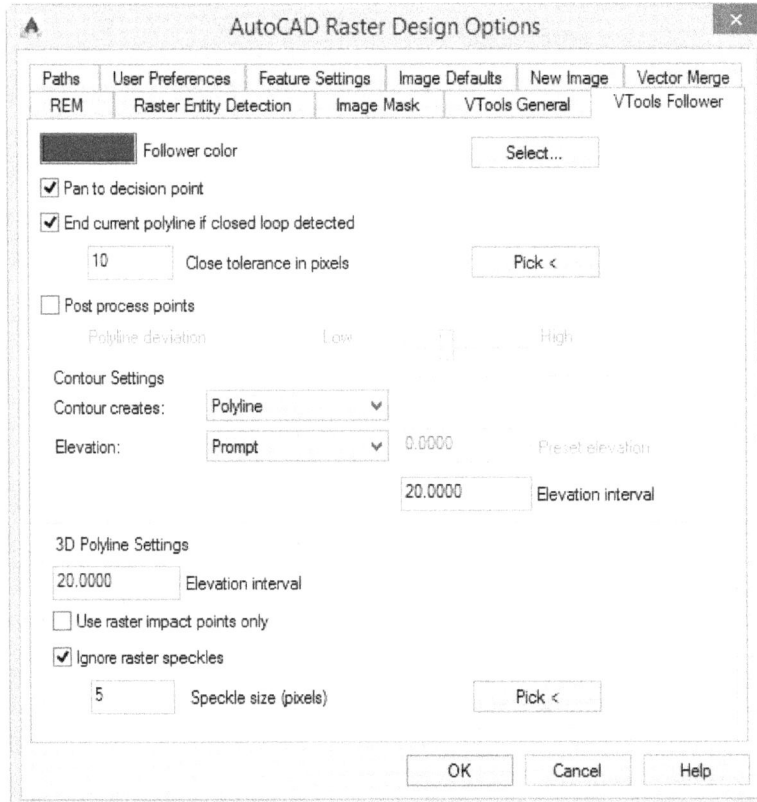

*Figure 6-2 The **AutoCAD Raster Design Options** dialog box with the **VTools Follower** tab chosen*

The **Follower color** button in this tab displays the current color assigned to the Followers tools which signifies the progress of the tool in the drawing. To change the color, you can either choose the **Follower color** or **Select** button in the **VTools Follower** tab. On choosing the button, the **Select Color** dialog box will be displayed. Choose the required color and then choose the **OK** button; the **Select Color** dialog box will be closed.

If you select the **End current polyline if closed loop detected** check box, it will automatically detect the endpoints of a polyline or a contour within the tolerance distance. Also, you can specify the distance between endpoints in the **Close tolerance in pixels** edit box available below this check box.

In the **Contour Settings** area, you can select the required option from the **Contour creates** and **Elevation** drop-down lists. The elevation value can be entered in the **Elevation interval** edit box.

In the **3D Polyline Settings** area, you can change values in the **Elevation interval** edit box to assign new elevation interval values. The **Use raster impact points only** and **Ignore raster speckles** check boxes control the new vector lines and elevation of raster contours, respectively. The speckle size can be entered in the **Speckle size (pixels)** edit box.

After applying general settings in the **Vtools General** and **VTools Follower** tabs; you can use the **Followers**, **Primitives**, **Text**, and **OCR** tools to convert raster entities in the drawing into vector entities. These tools are discussed next.

FOLLOWERS TOOLS

The Followers tools follow or trace the selected polyline and pause for input at any decision point where the follower can no longer proceed on its own. Different **Followers** tools are the **Polyline Follower**, **Contour Follower**, and **3D Polyline Follower**. These tools are discussed next.

Polyline Follower Tool

Ribbon:	Raster Tools > Vectorize & Recognize Text Panel > Followers drop-down > Polyline Follower
Command:	VFPLINE

The **Polyline Follower** tool traces the polyline and converts it into a vector entity. It continues tracing in the direction that seems most apparent to the follower. This option is useful in cases where the follower pauses at raster intersection and then again continues.

To trace a polyline and convert it into a vector entity, choose the **Polyline Follower** tool from the **Followers** drop-down in the **Vectorize & Recognize Text** panel of the **Raster Tools** tab. On doing so, the cursor will change into a crosshair and you will be prompted to specify a point on the raster entity; the entity will get converted into a vector object, as shown in Figure 6-3.

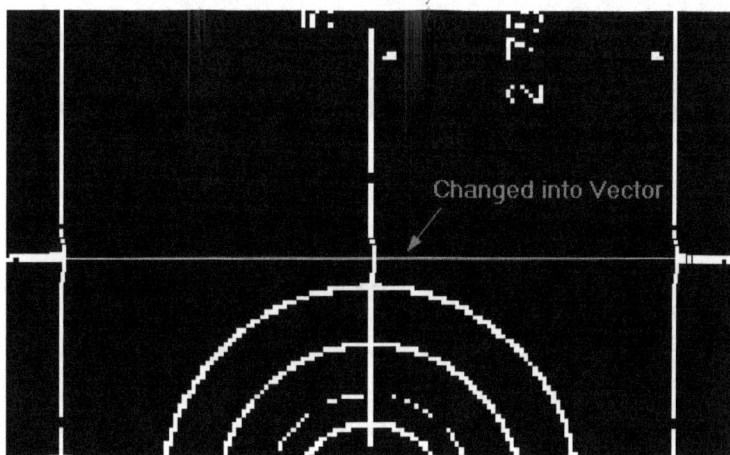

Figure 6-3 *Raster to vector conversion using the **Polyline Follower** tool*

Contour Follower Tool

Ribbon: Raster Tools > Vectorize & Recognize Text Panel > Followers drop-down >
                 Contour Follower
Command: VFCONTOUR

Sometimes you need to trace and digitize contour lines which is a tedious and time-consuming task. However, by using the **Contour Follower** tool, you can convert raster contour entities into vector entities quickly and easily.

To convert raster contour entities into vector entities, choose the **Contour Follower** tool from the **Followers** drop-down in the **Vectorize & Recognize Text** panel of the **Raster Tools** tab; you will be prompted to specify the point to follow. After selecting the contour, you will be prompted to specify elevation for contour. Enter the contour value and then press ENTER twice; the contour will change into vector entity as per the specified contour value, as shown in Figure 6-4.

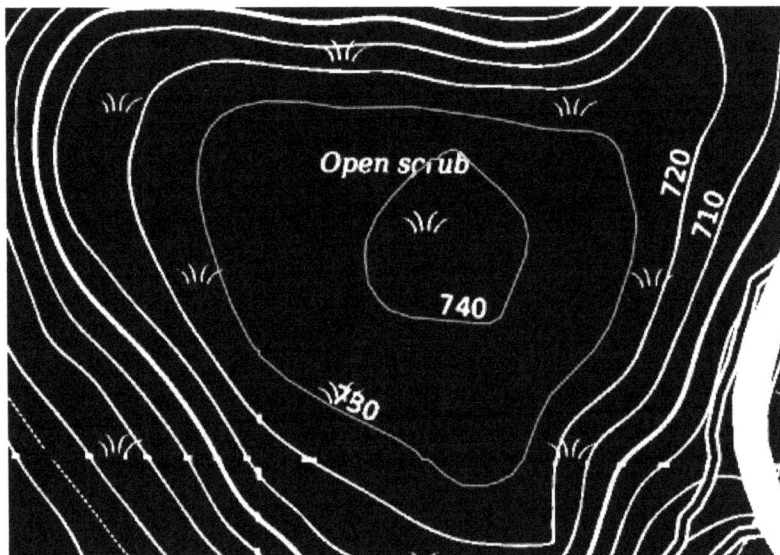

*Figure 6-4 Raster to Vector contour conversion using the **Contour Follower** tool*

3D Polyline Follower Tool

Ribbon: Raster Tools > Vectorize & Recognize Text Panel > Followers drop-down >
                 3D Polyline Follower
Command: VF3DPOLY

The **3D Polyline Follower** tool is used to create breaklines from the topo maps. It interacts with raster entities and the existing vectors to create 3D polylines. Using this tool, you can draw a polyline fence intersecting raster contours or you can select an existing vector polyline.

To create breaklines using this tool, choose the **3D Polyline Follower** tool from the **Followers** drop-down in the **Vectorize & Recognize Text** panel of the **Raster Tools** tab; you will be prompted to specify first fence point or follow vector. After specifying the first point, you will be again prompted to specify the end point. Click on the drawing to specify the end point and then right-click; a shortcut menu will be displayed, choose the **Enter** option. On doing so, you will be prompted to specify the elevation value. Specify the desired elevation value, a 3D breakline will be created, as shown in Figure 6-5.

Figure 6-5 *Creating 3D breakline using the* ***3D Polyline Follower*** *tool*

PRIMITIVES TOOLS

The Primitives tools convert raster data into vector data faster thereby eliminating the time required for cleanup steps. These tools give you more accurate results when you create lines and polylines from raster data. These tools work on bitonal images and are discussed next.

Line Tool

Ribbon:	Raster Tools > Vectorize & Recognize Text Panel > Primitives drop-down > Line
Command:	VLINE

The **Line** tool is used for converting a raster line into a vector line with its position and dimensions same as that of the selected raster line.

To create a vector line on the raster line, choose the **Line** tool from the **Primitives** drop-down in the **Vectorize & Recognize Text** panel of the **Raster Tools** tab; the cursor will change into a crosshair and you will be prompted to specify one pick point or two points in the drawing. Select the raster entity in the drawing; the selected raster entity will change into a line segment, as shown in Figure 6-6.

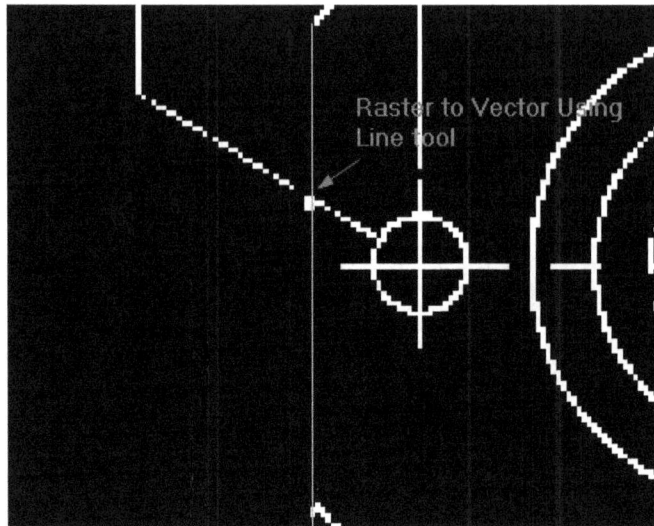

Figure 6-6 *Conversion of raster line entity into vector entity using the **Line** tool*

Polyline Tool

Ribbon: Raster Tools > Vectorize & Recognize Text Panel > Primitives drop-down >
                 Polyline
Command: VPLINE

The **Polyline** tool converts raster lines into vector polylines. It converts raster entities into vector entities with their position and dimensions same as that of the converted raster polyline. In this conversion, the selected raster entity is removed on exiting the command.

To convert a raster polyline into a vector polyline, choose the **Polyline** tool from the **Primitives** drop-down and then select the raster line to be converted; you will be prompted to select the **one-pick point**, **2P**, **Angle**, or **Length** option. Select any of the options to convert raster entity into vector. The raster line will be changed into vector polyline.

Rectangle Tool

Ribbon: Raster Tools > Vectorize & Recognize Text Panel > Primitives drop-down >
                 Rectangle
Command: VRECT

In AutoCAD Raster Design, you can convert raster rectangles into vector polyline by using the **Rectangle** tool. To convert a raster rectangle into a vector polyline, choose the **Rectangle** tool from the **Primitives** drop-down; the cursor will change into a crosshair and you will be prompted to specify the first corner point on the image. Click to specify the first point; you will be prompted to specify the angle for the line. Specify the second point to draw the rectangle. After creating the rectangle, you will be prompted to exit the command area. Choose the eXit option from the Command prompt area; the rectangle will be changed into vector polyline and background raster will be removed, as shown in Figure 6-7.

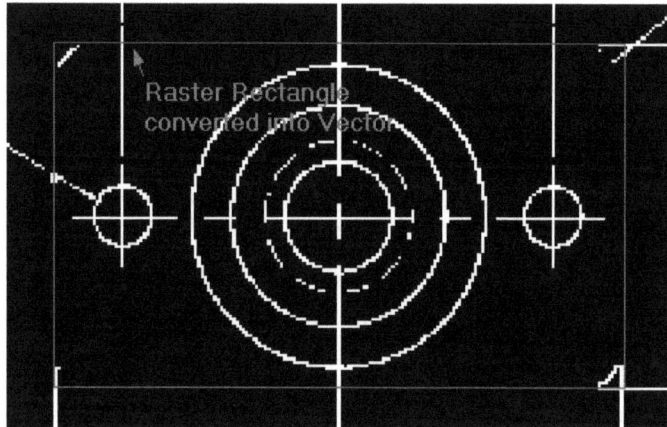

Figure 6-7 *Raster rectangle converted into vector polyline using the **Rectangle** tool*

Circle Tool

Ribbon:	Raster Tools > Vectorize & Recognize Text Panel > Primitives drop-down > Circle
Command:	VCIRCLE

The **Circle** tool enables you to convert a raster circle into a vector circle of the same size and geometry.

To convert a raster circle into vector circle, choose the **Circle** tool from the **Primitives** drop-down; the cursor will change into a crosshair and you will be prompted to pick a point on the raster circle. Now, click on the desired circle and press ENTER twice; the entity thus created will be a vector entity and the raster entity will be deleted, as shown in Figure 6-8.

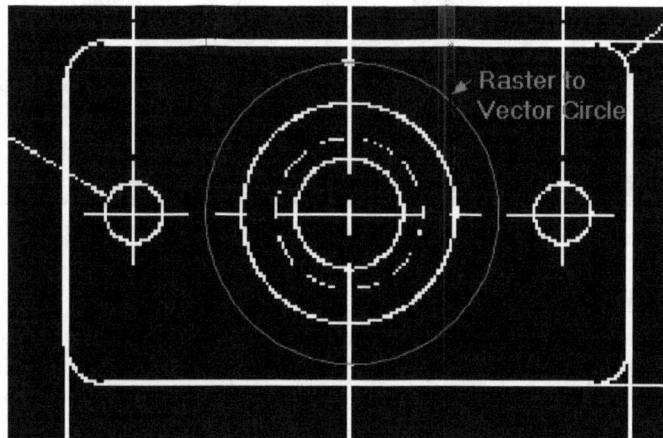

Figure 6-8 *Raster circle converted into vector circle using the **Circle** tool*

Note

*There are four methods of defining a raster circle: **One pick**, **Center**, **2P**, and **3P** multi-pick method.*

Arc Tool

Ribbon: Raster Tools > Vectorize & Recognize Text Panel > Primitives drop-down > Arc
Command: VARC

The **Arc** tool converts raster arc entities into vector arc entities of the same size and geometry.

To convert raster arc into vector entity, choose the **Arc** tool from the **Primitives** drop-down; the cursor will change into a crosshair and you will be prompted to pick a point on the raster arc. Click on the arc and press ENTER twice; the raster arc will be converted into a vector arc and the raster arc will be deleted, as shown in Figure 6-9.

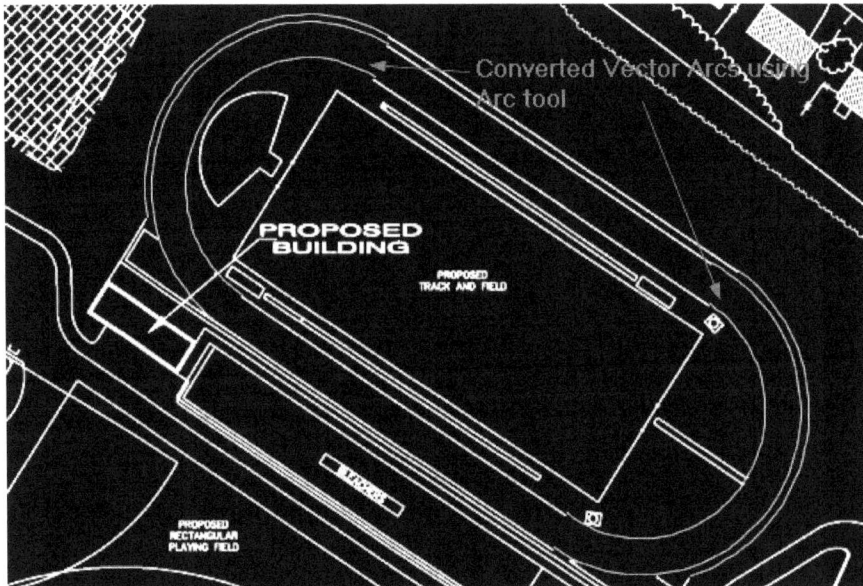

Figure 6-9 Raster arc converted into vector arc using the Arc tool

TEXT RECOGNITION TOOLS

AutoCAD Raster Design provides text recognition tools that work with bitonal raster images to recognize both machine-printed and hand-printed text and text in tables, and then converts it into AutoCAD text or mtext (multiline text) quickly and accurately. The results are displayed in the **Edit pane** highlighting words or characters that may not have been recognized accurately. Some of the important text recognition tools are discussed next.

Multiline Text Tool

Ribbon: Raster Tools > Vectorize & Recognize Text Panel > Text drop-down >
                 Multiline Text
Command: VMTEXT

The **Multiline Text** tool is used to create or edit raster text from the bitonal image. It creates a block of vector text around the selected area.

To edit or create a new entity, choose the **Multiline Text** tool from the **Text** drop-down in the **Vectorize & Recognize Text** panel of the **Raster Tools** tab; the cursor will change into a crosshair and you will be prompted to specify the first point on the desired area. To create or edit the raster text specify the first point in the desired area by drawing a rectangle. Then, write the desired text in the edit box and press ENTER; the text will be placed in the edit box, as shown in Figure 6-10.

Figure 6-10 Text overwritten using the **Multiline Text** tool

Note
*You can also edit the color, size, and text style of the text in the respective edit boxes of the **Text Editor** tab.*

Text Tool

Ribbon: Raster Tools > Vectorize & Recognize Text Panel > Text drop-down > Text
Command: VTEXT

Sometimes, you need to edit or create text in your existing raster drawings. Using AutoCAD Raster Design, you can create a vector text as a replacement for raster text or as a new entity. You can also select, edit, or move separate entities from the raster image.

To convert raster text into vector text, choose the **Text** tool from the **Text** drop-down in the **Vectorize & Recognize Text** panel of the **Raster Tools** tab; the cursor will change into a crosshair and you will be prompted to specify the insertion point. On selecting the insertion point, the **VText Edit** window will be displayed, as shown in Figure 6-11. Enter the text that you want to place in your image and choose the **OK** button; the text will be replaced, as shown in Figure 6-12. If you want to delete the underlying text, select the text and press ENTER; the text will be deleted from the image.

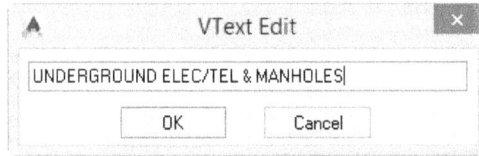

*Figure 6-11 The **VText Edit** window*

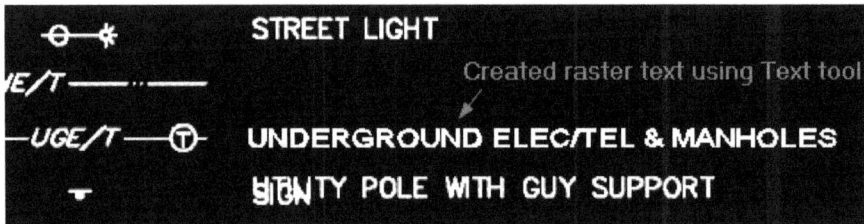

*Figure 6-12 Conversion of raster text into vector text using the **Text** tool*

Note
*After creating the raster text using the **Text** tool, you can edit the text properties from the **PROPERTIES** palette.*

OPTICAL CHARACTER RECOGNITION (OCR) TOOLS

Optical Character Recognition (OCR) tools help to recognize machine and hand-printed text and tables on the raster images and to create AutoCAD software text or multiline text (mtext). These tools also help to edit or manipulate raster texts or tables from the bitonal images. This interactive verification matches results with dictionary, speeds up manual data entry, and improves accuracy when converting drawings with text. The major OCR tools are discussed next.

Optical Character Recognition (OCR) Setup Tool

Ribbon:	Raster Tools > Vectorize & Recognize Text Panel > OCR drop-down > Optical Character Recognition (OCR) Setup
Command:	IRECSETUP

The **Optical Character Recognition (OCR) Setup** tool is used to adjust the settings for text.

To adjust the settings for text recognition, choose the **Optical Character Recognition (OCR) Setup** tool from the **OCR** drop-down in the **Vectorize & Recognize Text** panel of the **Raster Tools** tab; the **Text Recognition Setup** dialog box will be displayed, refer to Figure 6-13.

In this dialog box, you can specify text format, selection shape, dictionary for spellings, language character set, character types for recognition, output type, text height, removal method, text style, and display verification. These options are discussed next.

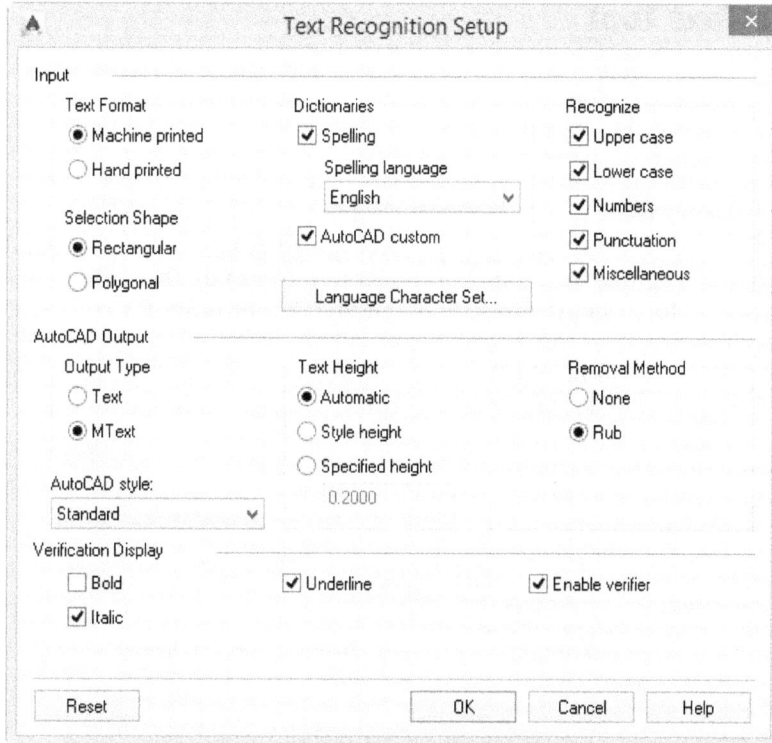

Figure 6-13 *The* **Text Recognition Setup** *dialog box*

In the **Input** area of the **Text Recognition Setup** dialog box, you can specify the settings for input text recognition operations. You can select the **Machine printed** or **Hand printed** radio button from the **Text Format** area. The **Machine printed** option detects the text if it has been created by a computer or other machine. The **Hand printed** option detects the text if it is hand written.

You can select the **Rectangular** or **Polygonal** radio button from the **Selection Shape** area of the **Text Recognition Setup** dialog box. This helps in detecting the rectangular or polygonal area when you place new text in the drawing.

In the **Dictionaries** area, you can check the spellings of the selected texts by selecting the **Spelling** and **AutoCAD custom** check boxes. If you select the **Spelling** check box, it will check all the spellings of the text as per dictionary standards during text recognition. You can also select the language of the desired text from the **Spelling language** drop-down list. If you select the **AutoCAD custom** check box, it will use AutoCAD user dictionary during spell check.

In the **Recognize** area, if you select the **Upper case** check boxe, the OCR tool will detect all the upper case alphabets in the raster image. And similar it goes for rest of the check boxes.

In the **AutoCAD Output** area, you can specify the **Output Type**, **Text Height**, **Removal Method**, and **AutoCAD style** during the text output process.

In the **Verification Display** area, you can select the **Bold**, **Underline**, **Enable verifier**, and **Italic** check boxes to display the selected format of text or table in the drawing.

Recognize Text Tool

Ribbon: Raster Tools > Vectorize & Recognize Text Panel > OCR drop-down >
                Recognize Text
Command: IRECTEXT

The **Recognize Text** tool helps to select raster text from the drawing and converts it into a single text or multitext. To convert raster text into vector text, choose the **Recognize Text** tool from the **OCR** drop-down in the **Vectorize & Recognize Text** panel of the **Raster Tools** tab; the cursor will change into a crosshair and you will be prompted to specify the first corner point for rectangle to recognize the text. On specifying the first corner point, you will be prompted to specify angle for text. On specifying the angle, you will be again prompted to specify second corner point to draw the rectangle; the **Verify Text** dialog box will be displayed, as shown in Figure 6-14.

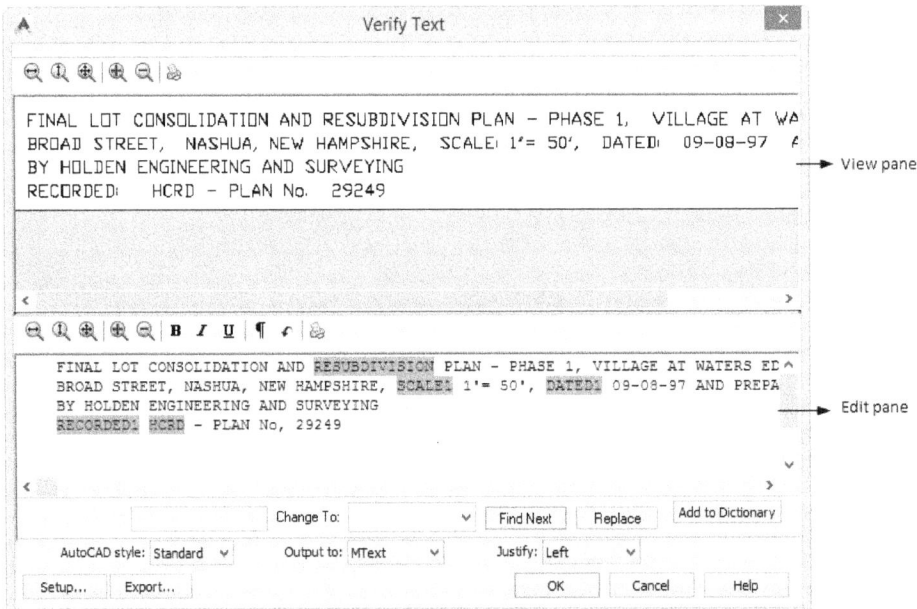

Figure 6-14 The **Verify Text** *dialog box*

The **Verify Text** dialog box is divided into two parts: the **View pane** and the **Edit pane**. The **View pane** displays the original raster text and the **Edit pane** displays the text in editable format. The incorrect words will be highlighted in green color in the **Edit pane** area. You can make corrections in the **Edit pane**. After all the corrections are done in the text, you can choose the **OK** button. The **Verify Text** dialog box is closed and the modified text will replace the text on the raster image.

Note
You can use various options in the ***Edit pane*** *to modify or edit raster text, such as zoom in, zoom out, text style, output format, and justify options.*

Recognize Table Tool

Ribbon:	Raster Tools > Vectorize & Recognize Text Panel > OCR drop-down > Recognize Table
Command:	IRECTABLE

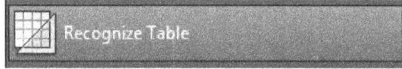

The **Recognize Table** tool recognizes the tabular text from the image and converts it into editable text.

To convert raster tabular text into editable vector text, choose the **Recognize Table** tool from the **OCR** drop-down in the **Vectorize & Recognize Text** panel of the **Raster Tools** tab; the cursor will change into a crosshair and you will be prompted to specify first corner point for the rectangle. On specifying the first corner point, you will be prompted to specify angle for text. On specifying the angle, you will be again prompted to specify second corner point to draw the rectangle; specify the second corner point. On doing so, the **Verify Table** dialog box will be displayed. You can modify the text in the **Edit pane** area. Now choose the **OK** button; the **Verify Table** dialog box will be closed and the table in the image will be modified.

TUTORIALS

Before starting the tutorial you need to download and save the tutorial files on your computer. To do so, follow the steps given next.

1. Download the *c06_rd_2016_tut.zip* file from *http://www.cadcim.com*. The path of the file is as follows:
   Textbooks > Civil/GIS > AutoCAD Raster Design > Exploring AutoCAD Raster Design 2016.

2. Now, save and extract the downloaded folder at the following location:

   C:\AutoCADRasterDesign2016

Notice that the *c06_rd_2016_tut* folder is created within the *AutoCADRasterDesign2016* folder.

Tutorial 1   Creating Basic Vector Objects

In this tutorial, you will convert the raster objects into vector objects using the **Line** tool and the **Arc** tool. **(Expected time: 30 min)**

The following steps are required to complete this tutorial:

a. Start AutoCAD Raster Design application and open the drawing file.
b. Convert raster lines into vector lines.
c. Convert raster lines into vector arcs.
d. Save the drawing file.

Starting AutoCAD Raster Design and Opening the Drawing File

1. Start AutoCAD Raster Design and choose the **Open** button from the Quick Access Toolbar; the **Select File** dialog box is displayed.

2. In this dialog box, browse to the following location:

    C:\AutoCADRasterDesign2016\ c06_rd_2016_tut

3. Select the **c06_Tut01.dwg** file from the *c06_rd_2016_tut* folder and then choose the **Open** button in the **Select File** dialog box; the geometry of the selected drawing is displayed in the drawing window.

Converting Raster Lines to Vector Lines
In this part, you will convert raster lines into vector lines using the **Line** tool.

1. Choose the **Area to be zoomed** option from the **Views** panel of the **View** tab; the required window will be visible, as shown in Figure 6-15.

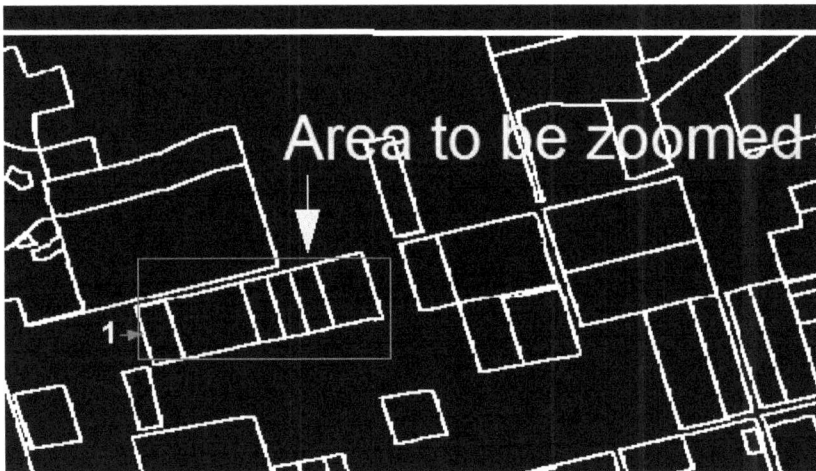

Figure 6-15 Area to be zoomed

2. Choose the **Line** tool from the **Primitives** drop-down in the **Vectorize &
Recognize Text** panel of the **Raster Tools** tab; the cursor is changed into a crosshair and you will be prompted to specify one-pick point on the object.

3. Click on the raster line marked as 1 on the raster image, refer to Figure 6-15; the line is converted into REM object. Next, press ENTER.

4. Similarly, click on all the raster lines shown in the marked box and then press ENTER twice; refer to Figure 6-15.

5. Ensure that the selected raster lines are converted into vector entities and all the raster lines under the new lines are removed, as shown in Figure 6-16.

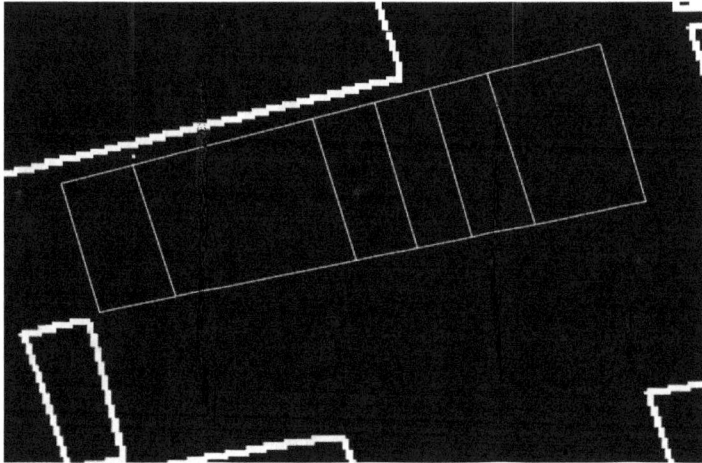

Figure 6-16 *Raster lines converted into vector lines*

Converting Raster to Vector Arcs

1. Choose the **Area to be converted** option from the **Views** panel of the **View** tab; the required view is visible in the drawing, as shown in Figure 6-17.

Figure 6-17 *Area to be converted*

2. Next, choose the **Arc** tool from the **Primitives** drop-down in the **Vectorize & Recognize Text** panel of the **Raster Tools** tab; the cursor is changed into a crosshair.

3. Select the raster arcs, refer to Figure 6-17 and ensure that rasters are changed into REM objects.

4. Press ENTER twice; the raster arcs are converted into vector arcs and underlying rasters are removed, as shown in Figure 6-18.

Figure 6-18 *Raster arcs converted into vector arcs*

Saving the Drawing File

1. Choose the **Save As > AutoCAD Drawing** from the Application Menu; the **Save Drawing As** dialog box is displayed.

2. In the **Save Drawing As** dialog box, enter the text **c06_Tut01_Result** in the **File name** edit box.

3. Choose the **Save** button; the **Save Image** dialog box is displayed.

4. Choose the **Save As** button from the **Save Image** dialog box; the **Save As** dialog box is displayed.

5. In this dialog box, browse to the following location:

    C:\AutoCADRasterDesign2016\c06_rd_2016_tut\c06_tut01

6. In the **File name** edit box, enter **r2v_Result**.

7. Ensure that the **Tagged Image File Format (*.tif, *.tiff)** option is selected from the **Files of types** drop-down list and then choose the **Save** button; the **Save As** dialog box is closed and the **Encoding Method** dialog box is displayed.

8. Ensure that the **Uncompressed** option is selected in the **Encoding** area and choose the **Next** button; the **Data Organization** dialog box is displayed.

9. Ensure that the **Stripped** option is selected in the **Organization** area and choose the **Finish** button; the raster image is exported to the specified location.

Closing the File

1. Choose the **Close** option from the Application Menu; the file is closed. Ignore the message box displayed, if any.

Tutorial 2   Creating Contours from Topo Maps

In this tutorial, you will convert a raster contour into a vector contour with defined elevation using the **Contour Follower** tool.   (**Expected time: 30 min**)

The following steps are required to complete this tutorial.

a. Start AutoCAD Raster Design application and open the drawing file.
b. Insert the image.
c. Convert raster contour into vector polyline.
d. Save the drawing.

Starting AutoCAD Raster Design and Opening the Drawing File

1. Start AutoCAD Raster Design application and then choose the **Open** button from the Application menu; the **Select File** dialog box is displayed.

2. In this dialog box, browse to the location *C:\AutoCADRasterDesign2016\c06_rd_2016_tut\ c06_tut02*, select the **c06_Tut02.dwg** file, and then choose the **Open** button; the file is opened in the drawing environment.

Inserting the Bitonal Image Using the Insert Tool

In this tutorial, you will insert grayscale image into your drawing for image enhancement.

1. Choose the **Insert** tool from the **Insert & Write** panel of the **Raster Tools** tab; the **Insert Image** dialog box is displayed.

    Insert...

2. In this dialog box, choose the **Quick insert** radio button from the **Insert Options** area.

3. Next, browse to the location *C:\AutoCADRasterDesign2016\c06_rd_2016_tut\c06_tut02* and then choose the **r2contour.tif** file.

4. Next, choose the **Open** button; the **Insert Image** dialog box is closed and the image is inserted in the drawing.

5. Next, enter **Z** in the Command prompt and press ENTER; you are prompted to specify the zoom option.

6. Enter **E** in the Command prompt and then press ENTER; the drawing zooms to its extents.

Converting Raster Contour into Vector Polyline

1. Zoom in to the specified area so that the contour (700) is clearly visible in the drawing window, refer to Figure 6-19, .

2. Choose the **Contour Follower** tool from the **Followers** drop-down of the **Vectorize & Recognize Text** panel of the **Raster Tools** tab; the cursor is changed into a crosshair.

3. Select the raster contour, refer to Figure 6-19; the REM line is created over the raster line.

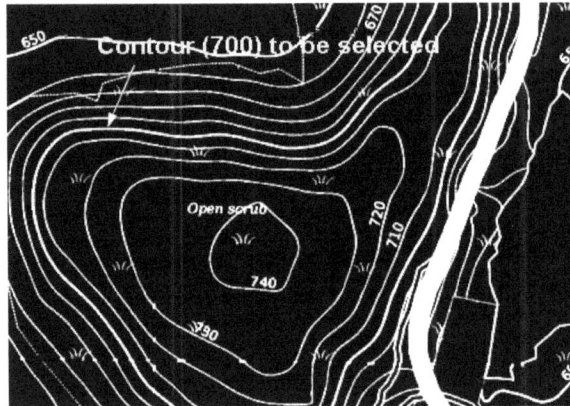

Figure 6-19 Raster polyline to be converted

4. Press ENTER; you are prompted to specify elevation for contour.

5. Enter **700** in the Command prompt area and then press ENTER twice; the contour line is created and the raster polyline is converted into vector polyline, as shown in Figure 6-20.

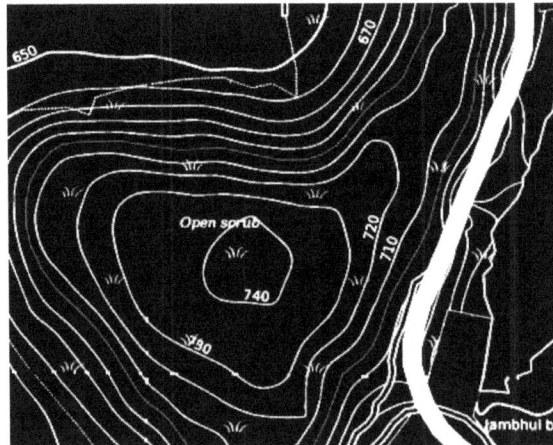

Figure 6-20 Converted contour line

Note
You can convert all the raster contours into vector by following the steps mentioned in this tutorial.

Saving the Drawing File

1. Choose **Save As > AutoCAD Drawing** from the Application Menu; the **Save Drawing As** dialog box is displayed.

2. In the **Save Drawing As** dialog box, enter **c06_Tut02_Result** in the **File name** edit box.

3. Choose the **Save** button; the **Save Image** dialog box is displayed.

4. Choose the **Save** button; the **Save Image** dialog box is closed and the drawing is saved at the specified location.

Closing the File

1. Choose the **Close** option from the Application Menu; the file is closed. Ignore the message box displayed, if any.

Tutorial 3   Editing Text Using Text Recognition Tool

In this tutorial, you will create new text as well as modify raster text in the drawing using the **Text** tool and the **OCR** (Optical Text Recognition) tool respectively.

(Expected time: 30 min)

The following steps are required to complete this tutorial:

a. Start AutoCAD Raster Design application and then open the drawing file.
b. Create a new text.
c. Edit existing raster text.
d. Save the drawing file.

Starting AutoCAD Raster Design and Opening the Drawing File

1. Start AutoCAD Raster Design and choose the **Open** button from the Quick Access Toolbar; the **Select File** dialog box is displayed.

2. In this dialog box, browse to the following location:

    C:\AutoCADRasterDesign2016\ c06_rd_2016_tut\c06_tut03

3. Select the **c06_Tut03.dwg** file from the *c06_tut03* folder and then choose the **Open** button; the geometry of the selected drawing is displayed in the drawing window.

Creating New Text

1. Zoom in to the specified location, as shown in Figure 6-21 so that the area is clearly visible in the drawing window.

2. Choose the **Text** tool from the **Text** drop-down in the **Vectorize & Recognize Text** panel of the **Raster Tools** tab; the cursor is changed into a crosshair.

3. Specify the insertion point in the drawing, refer to Figure 6-21; the **VText Edit** window is displayed.

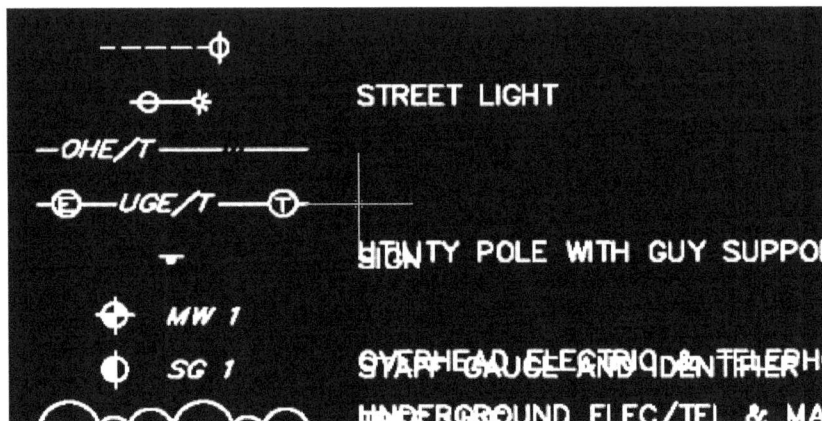

Figure 6-21 *Specifying insertion point in the drawing*

4. In the **VText Edit** window, enter **UNDERGROUND ELEC/TEL & MANHOLES** and choose the **OK** button; the **VText Edit** window is closed and the text is inserted in the drawing area.

5. Press the ESC key twice.

6. Select the text in the drawing and enter **PR** in the Command prompt area; the **PROPERTIES** palette is displayed.

7. In the **Text** area of the **PROPERTIES** palette, enter **0.10** and **0.50** in the **Height** and **Width factor** edit boxes, respectively.

8. Next, press ENTER; the text is modified according to the specified values, as shown in Figure 6-22. Close the **PROPERTIES** palette.

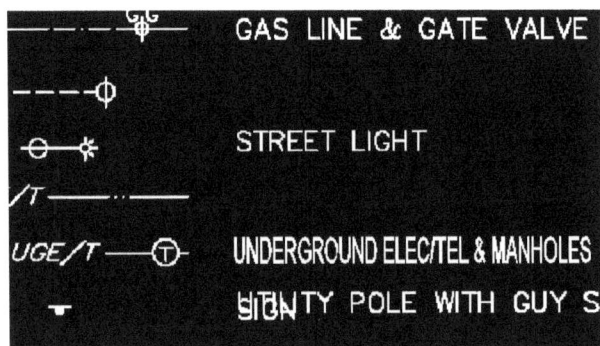

Figure 6-22 *The modified text*

Editing Existing Raster Text

1. Zoom in on the specified location so that the text is clearly visible in the drawing window, as shown in Figure 6-23.

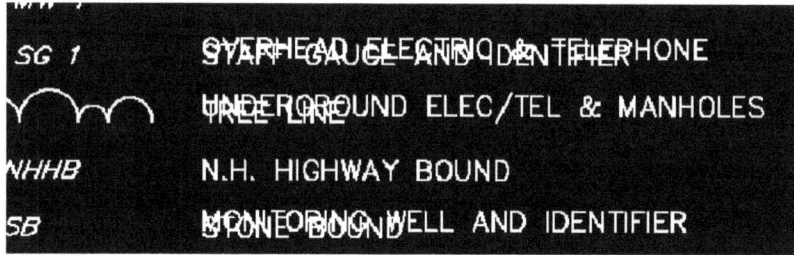

Figure 6-23 *The zoomed area*

2. Choose the **Recognize Text** tool from the **OCR** drop-down of the **Vectorize & Recognize Text** panel in the **Raster Tools** tab; the cursor is changed into a crosshair and you are prompted to specify the first point to draw a rectangle.

3. Click to specify the first corner point and then the second corner point. Next, move the cursor to the third corner point and click, refer to Figure 6-24; the **Verify Text** dialog box is displayed, as shown in Figure 6-25.

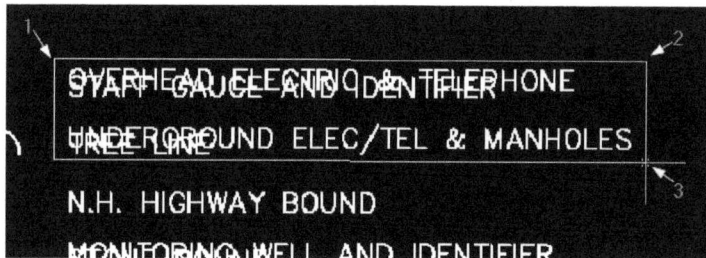

Figure 6-24 *Points to be selected*

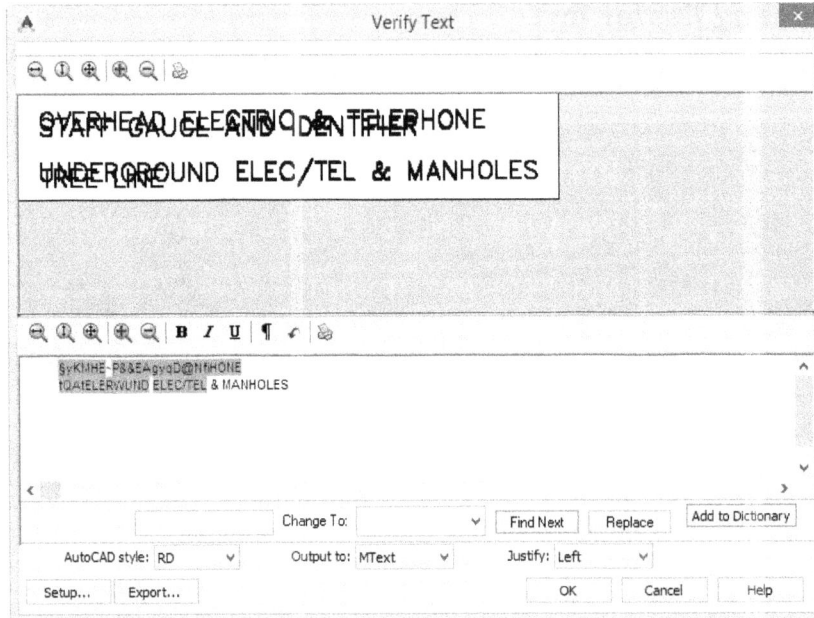

*Figure 6-25 The **Verify Text** dialog box*

4. In the **Edit pane** area of the **Verify Text** dialog box, delete the previous text and then enter the following text:

    STAFF GUARD
    TREE LINE

5. Next, choose the **OK** button; the **Verify Text** dialog box is closed and the drawing window is displayed with the edited text.

6. Now, press ESC twice to exit the command.

Note
*You can click in the Command prompt and press the ESC key if you do not want to continue with the **Recognize Text** tool.*

Saving the Drawing File

1. Choose **Save As** > **AutoCAD Drawing** from the Application menu; the **Save Drawing As** dialog box is displayed.

2. In the **Save Drawing As** dialog box, enter the text **c06_Tut03_Result** in the **File name** edit box.

3. Next, choose the **Save** button; the **Save Image** dialog box is displayed.

4. Choose the **Save** button; the **Save Image** dialog box is closed and the drawing is saved at the specified location.

Closing the File

1. Choose the **Close** option from the Application Menu; the file is closed. Ignore the message box displayed, if any.

Self-Evaluation Test

Answer the following questions and then compare them to those given at the end of this chapter:

1. Which of the following options is used to convert raster polyline into vector contour?

    (a) **3D Polyline Follower**   (b) **Contour Follower**
    (c) **Polyline Follower**   (d) All of the above

2. Which of the following methods is used to create a vector line from a raster line?

    (a) **Line Primitive**   (b) **Arc Primitive**
    (c) **Polyline Primitive**   (d) None of these

3. Which of the following tools is used to convert a vector circle into a raster circle?

    (a) **Arc Primitive**   (b) **Line Primitive**
    (c) **Circle Primitive**   (d) All of the above

4. Which of the following methods is used to remove underline raster entities?

    (a) **None**   (b) **Rub**
    (c) **REM**   (d) None of these

5. Which of the following methods is used to create single vector text from raster text?

    (a) **Multiline Text**   (b) **Text**
    (c) **Recognize Text**   (d) None of these

6. To create 3D polylines from raster maps, you need to choose the _____ tool from the **Follower** drop-down.

7. You can specify elevation value for contour creation in the _____ area.

8. You can specify a range of line widths for the raster entity in the **General** tab of the _____ dialog box.

9. To make the most efficient use of the **Polyline Follower**, you should adjust the settings of the _____ tab in the **AutoCAD Raster Design Options** dialog box.

10. You can use the **Text** tool to create a line of AutoCAD vector text as a replacement for raster text or as a new entity. (T/F)

11. To make the most efficient use of the vectorization tools, you should adjust the settings on the **VTools Follower** tab of the **Raster Design Options** dialog box. (T/F)

12. The **Contour Follower** tool traces raster contour lines and converts them into AutoCAD polyline entities. (T/F)

Review Questions

Answer the following questions:

1. The **Multiline Text** tool is used to convert raster text into _____ text.

2. The **3P** multi-pick method converts a raster arc by selecting _____ points along the arc.

3. The **PR** command in the Command prompt area is used to show the _____.

4. The _____ tool is used to convert raster text into vector text.

5. The _____ tool is used to vectorize a contour.

6. You can use the _____ tools to select raster text in your drawing and convert it into AutoCAD text or multiline text (**Mtext**).

7. To modify search parameters in the **Text Recognition Setup** dialog box choose the **Setup** button in the **Edit pane**. (T/F)

8. To make the most efficient use of the **Rectangle** tool, you should first adjust the settings on the **VTools General** tab in the **Raster Design Options** dialog box. (T/F)

9. The **Polyline Follower** tool traces vector polylines and converts them into AutoCAD polyline entities. (T/F)

10. Vectorizing tools work only on grayscale images. (T/F)

11. The **View pane** (upper pane) displays the original raster text in the **Verify Text** dialog box. (T/F)

12. The **3D Polyline Follower** tool considers the nearest point as the start point while following an existing polyline. (T/F)

EXERCISES

Exercise 1

Download the *c06_rd_2016_exr.zip* from *http://www.cadcim.com* and zoom in to the specified location for the given image, as shown in Figure 6-26. Next, convert the Proposed Softball, Baseball, and rectangular playing field raster objects into vector. **(Expected time: 40 min)**

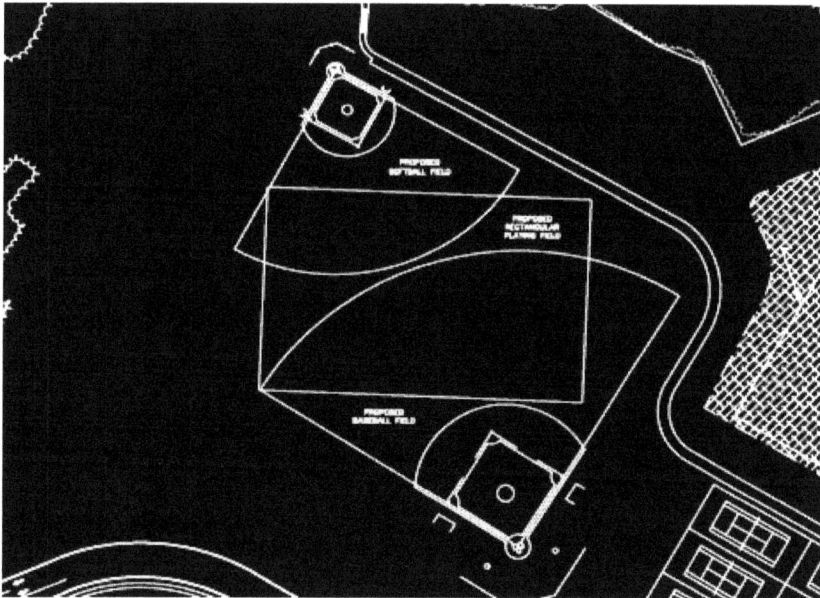

Figure 6-26 Rasters to be converted into vector

Exercise 2

Download the *c06_rd_2016_exr.zip* from *http://www.cadcim.com* and add **OVERHEAD ELECTRIC & TELEPHONE** text for the given image in the cursor location, refer to Figure 6-27.

(Expected time: 30 min)

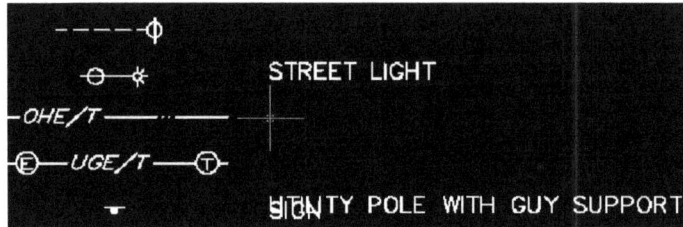

Figure 6-27 *Area in which text needs to be filled*

Chapter 7

Multispectral Images and Digital Elevation Models

Learning Objectives

After completing this chapter, you will be able to:
- *Understand concepts of Satellite Remote Sensing*
- *Understand concepts of Multispectral Remote Sensing*
- *Use Digital Elevation Models*

INTRODUCTION

AutoCAD Raster Design has a complete suite of tools for working with scanned maps, aerial photographs, and other digital raster images. You can use AutoCAD Raster Design tools to insert, edit, correlate, convert, and manage your images quickly and easily, regardless of the source of retrieval.

This chapter discusses about the raster images and their specifications. These raster images help in object identification. In this chapter, you will learn to open multispectral and hyperspectral satellite images such as Landsat MSS, TM, ETM, IRS P5, IRS P6, Hyperion, and so on. You will also learn to change band combination of a satellite image.

Moreover, you will learn about the Digital Elevation Models (DEM) and method to apply the hillshade effects on them. Additionally, you will be able to create slope map and aspect map using these digital elevation models.

SATELLITE REMOTE SENSING

The term "remote sensing" was first used in the United States in 1950s. It is now commonly used to describe the science and art of identifying, observing, and measuring an object without any direct contact with it. This process involves detection and measurement of radiation of different wavelengths reflected or emitted from distant objects or materials. In order to study large areas of the Earth's surface, devices known as sensors are used. These sensors are mounted on platforms such as helicopters, planes, and satellites that make it possible for the sensors to observe the Earth from distant heights.

Remote sensing sensor detects information reflected or emitted from the earth objects and stores coded information in the sensors. Then, remote sensing data analyst decodes this coded information into digital image.

Remote sensing images are normally in the form of digital images. These images are composed of pixels. Remote sensing technique allows sensors to capture images of the earth surface at different wavelength region of the Electromagnetic Spectrum (EMS). Some of the images depict solar radiation reflected in the visible and the near infrared regions of the electromagnetic spectrum; others are measurements of the energy emitted by the earth surface itself such as in the thermal infrared wavelength region. The energy measured in the microwave region is the measure of relative return from the earth's surface, where energy is transmitted from the satellite itself. This is known as active remote sensing, since the energy source is provided by the remote sensing platform. Whereas, the systems where remote sensing measurements depend upon the external energy source such as Sun, are referred as passive remote sensing systems.

Remote sensing technology began with the invention of the camera around 150 years back. The first aerial photographs were taken as "still" using balloons for topographic mapping. By the First World War, cameras mounted on airplanes provided aerial views of fairly large surface areas that proved invaluable in military reconnaissance. From then until the early 1960s, the aerial photographs remained the single standard tool for depicting the surface. Then multispectral remote sensing was developed for resource monitoring. Hyperspectral imaging is used to obtain information from the spectrum for each pixel in the image of a scene with the purpose of identifying manmade materials, detection of minerals and ores, and so on.

Remote sensing imagery has many applications in mapping land use and land cover (LU/LC), agriculture, soil, forestry, city planning, archaeological investigations, military observations, geomorphological surveying, land cover changes, deforestation, vegetation dynamics, water quality dynamics, urban growth, and so on.

Principles of Remote Sensing

For sustainable development, remote sensing process involves the collection, processing, and analysis of data for decision making. Remote sensing process includes detecting and recording of radiant energy reflected or emitted by objects or surface material, as shown in Figure 7-1. Each object reflects or emits electromagnetic radiation. The space mounted remote sensing sensors detect this reflected or emitted energy in the form of DN values.

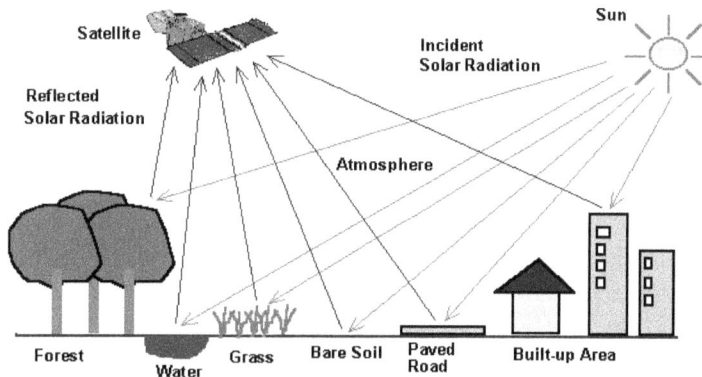

Figure 7-1 *Remote sensing process*

Remote sensing is basically a multi-disciplinary science which includes a combination of various disciplines such as optics, spectroscopy, photography, computer, electronics and telecommunication, satellite launch, and so on. All these technologies are integrated into a common system known as remote sensing. It comprises of various phases which are discussed next.

1. Emission of electromagnetic radiation or EMR (sun/self-emission).

2. Transmission of energy from the source to the surface of the earth as well as absorption and scattering.

3. Interaction of EMR with the earth's surface: reflection and emission.

4. Transmission of energy from the surface to the remote sensor.

5. Generation of sensor data output.

6. Data transmission, processing, and analysis.

Electromagnetic Radiation and Electromagnetic Spectrum

Electromagnetic Radiation (EMR) is a dynamic form of energy that propagates as wave motion at the speed of light c = 3 x 10^{10} cm/sec. The parameters that characterize a wave motion are wavelength (λ), frequency (v), and velocity (c), as shown in Figure 7-2. The relationship between the above is velocity = wavelength * frequency.

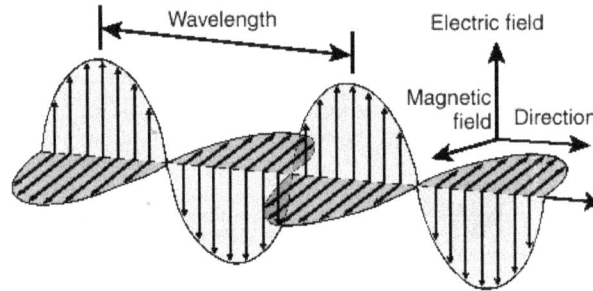

Figure 7-2   *Electromagnetic wave and its components*

The electric field and the magnetic field are important concepts that can be used to mathematically describe the physical nature of electromagnetic waves (light). The electric field and magnetic fields both are perpendicular to each other with respect to the direction of electromagnetic wave, refer to Figure 7-2. These two fields oscillate in a consistent manner so that the wave moves forward at a constant rate in accordance to the speed of light (c).

Electromagnetic (EM) energy radiates in accordance to the basic wave theory. This theory describes that the EM energy travels in a harmonic sinusoidal fashion at the velocity of light. Many characteristics of EM energy are easily described by wave theory whereas particle theory offers insight into how electromagnetic energy interacts with matter.

The electromagnetic spectrum is the term used by the scientists to describe the entire range of existing light. The electromagnetic spectrum describes all wavelengths of light. Light is a wave of altering electric and magnetic fields, as shown in Figure 7-3. The propagation of light is similar to the waves crossing an ocean. The larger the frequency, the smaller will be the wavelength and vice-versa.

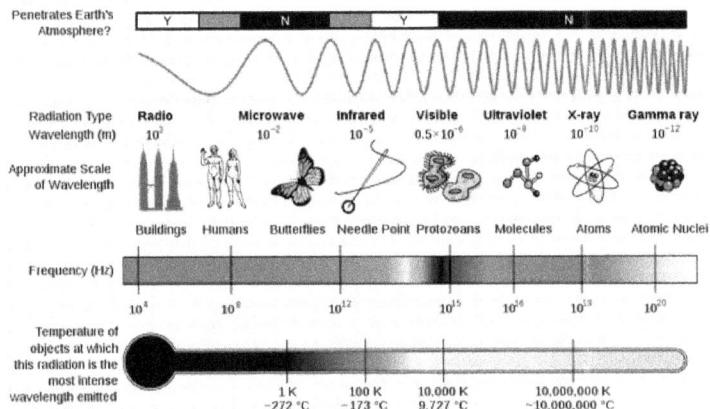

Figure 7-3   *Electromagnetic Spectrum*

Our eyes detect only visible light that oscillates between 400 and 790 terahertz (THz). This is several hundred trillion times of a second. The wavelength ranges from 390 – 750 nanometers (1 nanometer = 1 billionth of a meter). Our brain interprets various wavelengths of light in different colors. Red has the longest wavelength and the violet has the shortest. When sunlight passes through a prism, you will see that it is actually composed of many wavelengths of light. The prism creates a rainbow by redirecting each wavelength out at a slightly different angle.

In general, the longer wavelengths come from the coolest and darkest regions of space. Meanwhile, the shorter wavelengths indicate extremely energetic phenomena.

Wavelengths shorter than violet are ultraviolet or UV lights. Our atmosphere blocks the UV light. Beyond UV, the highest energies in the electromagnetic spectrum are the X-rays and the gamma rays. X-rays are generated from neutron stars and gamma rays are the shortest wavelengths of light.

Types of Remote Sensing

The Sun is the primary source of energy for applying remote sensing technology. Based on the source of energy, the process of acquiring image can be classified into active and passive remote sensing, refer to Figure 7-4.

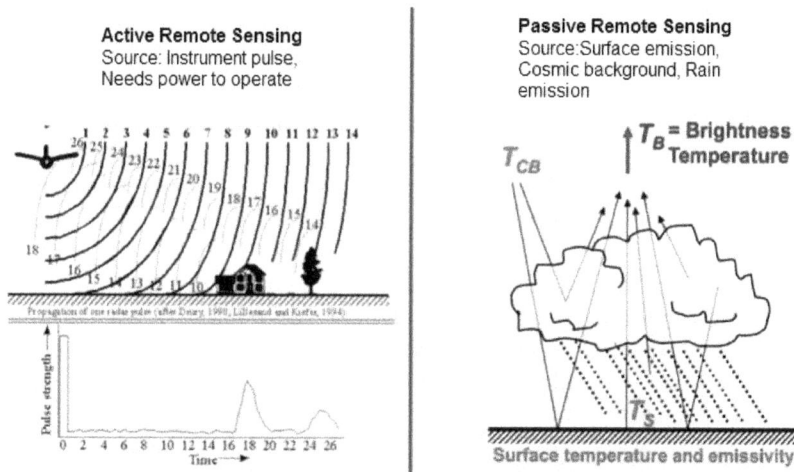

Figure 7-4 *Active and passive remote sensing process*

Remote sensing systems that send their own source of energy for illumination are known as Active sensors. These sensors have the advantage of acquiring data any time of the day or the season. Synthetic Aperture Radar (SAR) is a good example of active sensor. Remote sensing systems which detect the natural available energy and then retrieve information from it are called Passive sensors. However, passive sensors can collect information only in day light. Energy that is naturally emitted or reflected can be detected both in day and night provided that the amount of energy is large enough to be recorded.

On the basis of electromagnetic spectrum, remote sensing technology can also be classified into three types: optical, thermal, and microwave remote sensing.

The optical remote sensing technology deals with visible, near infrared, middle infrared, and short wave infrared regions of the EMS. Most of the remote sensors record the EMR in the range of 300 nm to 3000 nm, such as bands of IRS P6 LISS IV sensors are in optical range of EMR.

Thermal remote sensing sensors deal in thermal range of EMS. These sensors record the energy from the wavelength range of 3000 nm to 5000 nm and 8000 nm to 14000 nm. They can also collect data even for high temperature phenomenon such as forest fire. For example, Band 6 of Landsat ETM+ collects thermal data.

Microwave remote sensing collects spatial data from 1 mm to 1 m range of electromagnetic spectrum. Most of the microwave sensors are active sensors with their own source of light. These sensors are weather independent and can collect data at night. Synthetic Aperture Radar (SAR) is an example of microwave remote sensing.

Image Resolutions

Resolution term can be defined as the ability to separate and distinguish adjacent objects or items in a scene irrespective of the color, size, tone, and brightness of the adjacent entities. Remote sensors measure differences and variations of objects that are often described in terms of four main resolutions: spatial, spectral, radiometric, and temporal which are discussed next.

The spatial resolution is the measure of the smallest spatial object on the Earth made by a sensor. It represents the area covered by a pixel on the ground. The smaller the spatial resolution, the greater will be the resolving power of the sensor. For example, IKONOS AVHRR sensor (spatial resolution 1*1 m) has better resolution than IRS P6 LISS III sensor.

The spectral resolution refers to the band width of the sensor. These bands can record data in the specific wavelength interval of the electromagnetic spectrum. The increase in the interval represents coarse spectral resolution and narrow interval represents fine spectral resolution. IRS OCM sensor has 8 narrow spectral bands thus resulting in high resolution imagery compared to IRS LISS IV with 3 spectral bands.

The radiometric resolution refers to the number of gray levels measured between pure black to pure white. Radiometric resolution is also measured by number of bits into which the recorded energy is divided. For instance, radiometric resolution of ASTER data is 8-bits for its first nine bands whereas for LISS III radiometric resolution is 7-bits.

Satellite images captured of a scene on different days help to detect the changes in a particular area. It helps to provide information about the changing of the variables through time. The temporal resolution refers to its revisit period. CARTOSAT-1 can capture images of the same area in every 5 days whereas temporal resolution of LISS III captures in 24 days.

MULTISPECTRAL REMOTE SENSING

Multispectral image is a set of images representing the same area at different wavelengths of the electromagnetic spectrum. Satellite sensors detect and capture the reflected wavelengths at different bands. Each band of a multispectral image set is considered as an image from a different segment of the electromagnetic spectrum. You can set the color combination of an image by changing its band combination.

In AutoCAD Raster Design, you can work with multispectral and hyperspectral data. The multispectral data gives information of the entire earth surface. By changing band combinations, you can identify entities of the inaccessible areas. You can also create false color images to identify surface features easily.

Band Combination

Remote sensing images from satellite as well as aircraft are highly useful for facility design, construction, and maintenance. In AutoCAD Raster Design, you can combine several data bands to highlight specific spatial object. Once a multispectral image is inserted in your drawing, you can change or edit the band combination of the multispectral image. By analyzing different combinations of color bands, you can identify spatial objects which are not visible in the image. Landsat MSS data is available in four bands. On changing the band ratio, you can assign natural color, false color, and standard false color to the image. Satellite sensors measure electromagnetic radiation in different areas of electromagnetic spectrum.

For instance, to create a standard false color image of a Landsat MSS data, you need to pass Band 3 in Red channel, Band 2 in Green channel, and Band 1 in Blue channel. Further details about the Landsat MSS image are discussed next.

Landsat Multispectral Scanner (MSS) images consist of four spectral bands with 60 meter spatial resolution. In these images, approximate scene size is 170 km in north-south direction and 185 km in east-west direction. Band designations differ from Landsat 1-3 to Landsat 4-5 which are shown in Table 7-1.

Table 7-1 Band designations for the Landsat MSS sensor

Multispectral Scanner (MSS)	Landsat 1-3	Landsat 4-5	Wavelength (micrometers)	Resolution (meters)
	Band 4	Band 1	0.5-0.6	60
	Band 5	Band 2	0.6-0.7	60
	Band 6	Band 3	0.7-0.8	60
	Band 7	Band 4	0.8-1.1	60

AutoCAD Raster Design provides options and tools to insert and edit a multispectral image to make it suitable for your research. These tools are discussed next.

Inserting Multiband Image

Ribbon: Raster Tools > Insert & Write > Insert
Command: IINSERT

To open different bands of an image from a multispectral dataset, choose the **Insert** tool from the **Insert & Write** panel of the **Raster Tools** tab; the **Insert Image** dialog box will be displayed, refer to Figure 7-5. The options in the **Insert Image** dialog box have already been discussed in Chapter-2.

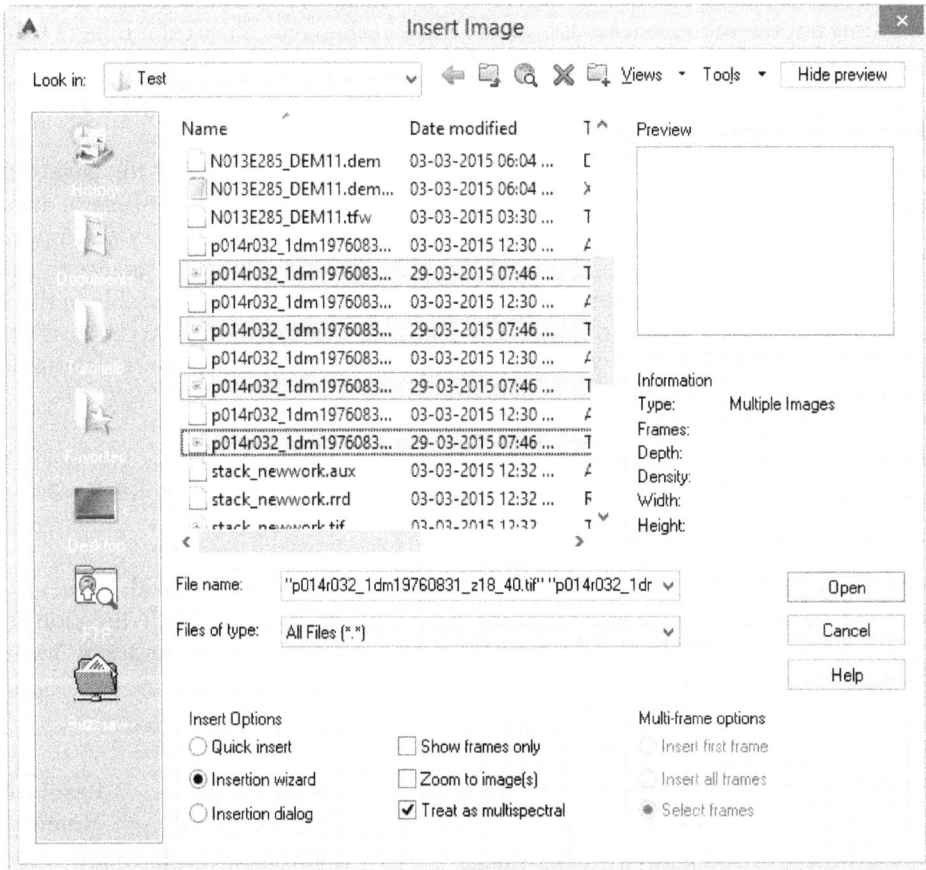

*Figure 7-5 The **Insert Image** dialog box with multiple bands selected*

Select the desired image layers from the **Insert Image** dialog box, refer to Figure 7-5. Also, ensure that the **Treat as multispectral** check box and the **Insertion wizard** radio button are selected. Next, choose the **Open** button; the **Assign Color Map** page will be displayed, as shown in Figure 7-6. The details about these pages are discussed next.

Note
*You need to select multiple *.tiff format images from the **Insert Image** dialog box to activate the* ***Treat as multispectral*** *check box.*

In the **Assign Color Map** page, you can select the desired band combination by selecting the image from the drop-down list corresponding to the **Red**, **Green**, and **Blue** check boxes. In this page, you can also create false color of an image by assigning infrared band in the red channel, red band in the green channel, and green band in the blue channel. Then, select the **Insert into display** check box and choose the **Next** button. The **Pick Correlation Source** page will be displayed, as shown in Figure 7-7.

Figure 7-6 *The Assign Color Map page* 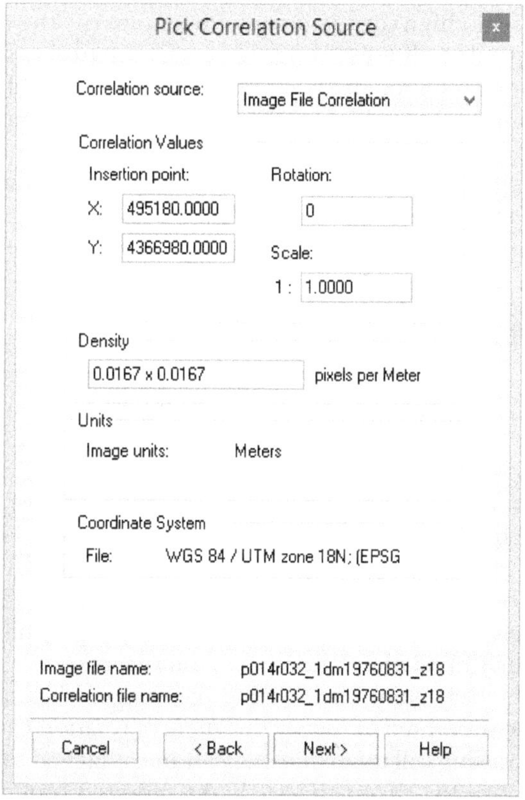   *Figure 7-7* *The Pick Correlation Source page*

In the **Pick Correlation Source** page, you can specify the source for image correlation by selecting an option from the **Correlation source** drop-down list. The options in the **Correlation Values** area display the insertion point coordinates, scale, and rotation of the image. The insertion point coordinates can be specified in the **X** and **Y** edit boxes. Similarly, the **Scale** and the **Rotation** of the image can be defined in the **Scale** and **Rotation** edit boxes, respectively. The values displayed in the **Correlation Values** area are read from the source which is specified in the **Correlation source** drop-down list.

The **Density** edit box in the **Pick Correlation Source** page displays the pixel density of the raster per image unit. The image unit is displayed in the **Units** area below the **Density** edit box, refer to Figure 7-7. The **Coordinate System** area in this page displays the Coordinate Reference System (CRS) of the selected raster image.

After specifying the required options in this page, choose the **Next** button; the **Modify Correlation Values** page will be displayed. By default, the options in the **Modify Correlation Values** page will display the same correlation values as were specified in the **Pick Correlation Source** page.

You can use the options in the **Modify Correlation Values** page to modify the correlation parameters for the image to be inserted, as shown in Figure 7-8.

To modify a correlation parameter, specify the required value in the **Correlation Values** area. To specify a different image unit, select an option from the **Image units** drop-down list in the **Units** area. Additionally, you can specify the required scale value in the **Scale** edit box. To enlarge the image, you can specify the scale value between 0 to 1.

After modifying the required correlation parameters, choose the **Apply** button; the modified correlation values will be applied to the image. Choose the **Next** button; the **Insertion** page will be displayed.

Note
*While inserting an image with a coordinate system different than the one assigned to the drawing, the **Transform** page of the **Insertion** wizard will be displayed in succession to the **Modify Correlation Values** page. You can use the options in the **Transform** page to transform the coordinate system of the image.*

In the **Insertion** page, you can graphically or numerically specify the correlation parameters required for inserting an image, as shown in Figure 7-9.

To specify the parameters numerically, enter the values in the required edit boxes as explained in the previous section. Alternatively, to graphically specify the values, choose the **Pick** button in the **Correlation Values** area; the **Insertion** wizard will close and you will be prompted to specify the base point for inserting the image. Click on the required location in the drawing; you will be prompted to specify the angle of rotation of the image. Notice that, moving the cursor in the drawing will rotate the image frame attached to the cursor according to the specified base point. Move the cursor to specify the required frame size and then click; the image will be inserted into your drawing and the **Insertion** wizard will be displayed. Choose the **Finish** button in the **Insertion** wizard; the image is inserted in the drawing based on the correlation parameters specified, as shown in Figure 7-10.

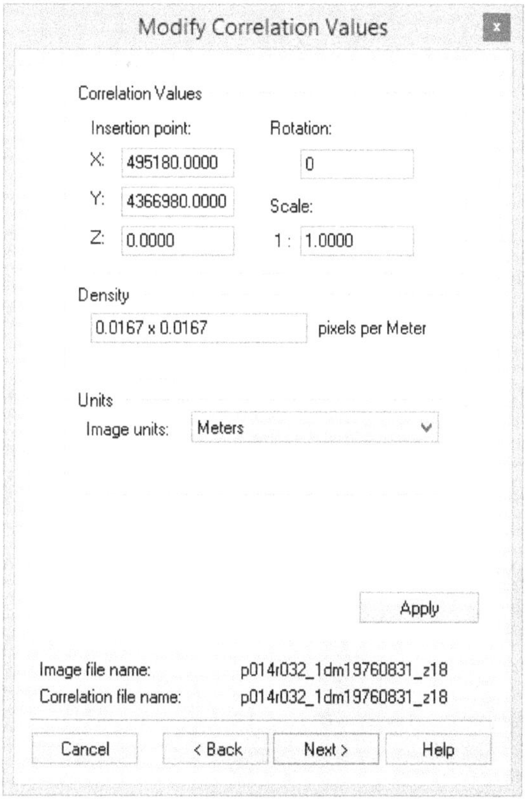

*Figure 7-8 The **Modify Correlation Values** page*

*Figure 7-9 The **Insertion** page*

Figure 7-10 Image inserted in the drawing

Note

*If the image is not displayed in the drawing window, enter **Z** in the Command prompt area and press ENTER. Then select the **Extent** option or press **E** in the Command prompt area; the image will be displayed in the drawing.*

Tip

*You can insert multiple band images in the drawing and you can also change the band combination of the image such as natural color, false color composition (FCC), or standard false color composition by changing the band combination in the **Assign Color Map** page.*

Editing Band Combination

After inserting the Landsat MSS data in the drawing, you can create false color images. To create false color band combination of a Landsat MSS image, right-click on the inserted image in the **RASTER DESIGN** palette; a shortcut menu will be displayed, as shown in Figure 7-11. Notice that the **Image Insertions** option is selected by default in the Image View drop-down list of the **RASTER DESIGN** palette.

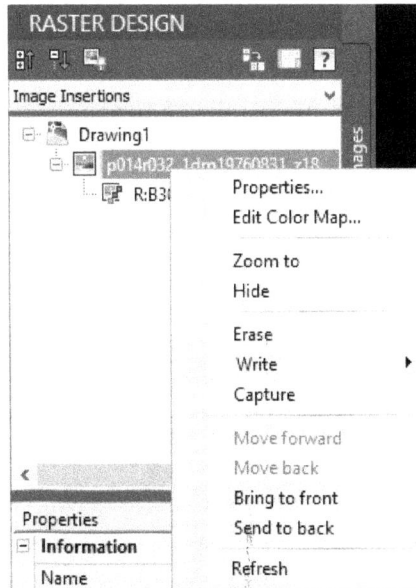

*Figure 7-11 Shortcut menu displayed in the **Image Insertions** view of the **RASTER DESIGN** palette*

On choosing the **Edit Color Map** option from the shortcut menu, the **Band Assignment Color Map** dialog box will be displayed, as shown in Figure 7-12. You can assign different band combinations to assign false color to the inserted multispectral image.

Figure 7-12 *Bands specified in the* ***Band Assignment Color Map*** *dialog box*

To assign false color to an image or edit a color map for a multispectral image, you can specify options in the **Band Assignment Color Map** dialog box. Click on the drop-down list corresponding to the **Red**, **Green**, and **Blue** check boxes to assign the desired band ratio. If you clear the check box corresponding to the desired channel, the desired color channel will be turned off. After specifying desired band combination, choose the **OK** button. The **Band Assignment Color Map** dialog box will be closed and desired band combination image will be displayed in the drawing, as shown in Figure 7-13.

Figure 7-13 *False color image created*

DIGITAL ELEVATION MODELS

A digital elevation model provides information of the elevation of a land surface. DEM files are represented as *.dem*. These DEM files are composed of point elevation data. DEM files are used to store and transfer large-scale topographic relief information. These DEM files are used to represent 3D information of the land surface and can be used in resource management, land planning, surveying, and engineering projects.

The DEM data is ideal for large scale planning and analysis; but not precise for small-scale studies. DEM files are valuable data source for planning and engineering projects. Information extracted from DEM is widely used for hydrological studies, corridor planning, landuse planning, slope analysis, and so on. AutoCAD Raster Design provides options and tools to edit digital elevation models to make it suitable for the desired research which are discussed next.

Inserting Digital Elevation Model

Ribbon: Raster Tools > Insert & Write > Insert
Command: IINSERT

Digital elevation models are used in three dimensional generation of slope, aspect, and terrain profile. These models are 3D representation of earth surface and are used to generate slope, aspect, and hillshade map of a particular region.

To open a digital elevation model in your drawing, choose the **Insert** tool from the **Insert & Write** panel of the **Raster Tools** tab; the **Insert Image** dialog box will be displayed, as shown in Figure 7-14. The options in the **Insert Image** dialog box have already been discussed earlier in Chapter 2. Ensure that the **Insertion wizard** option is selected in the **Insert options** area of the **Insert Image** dialog box.

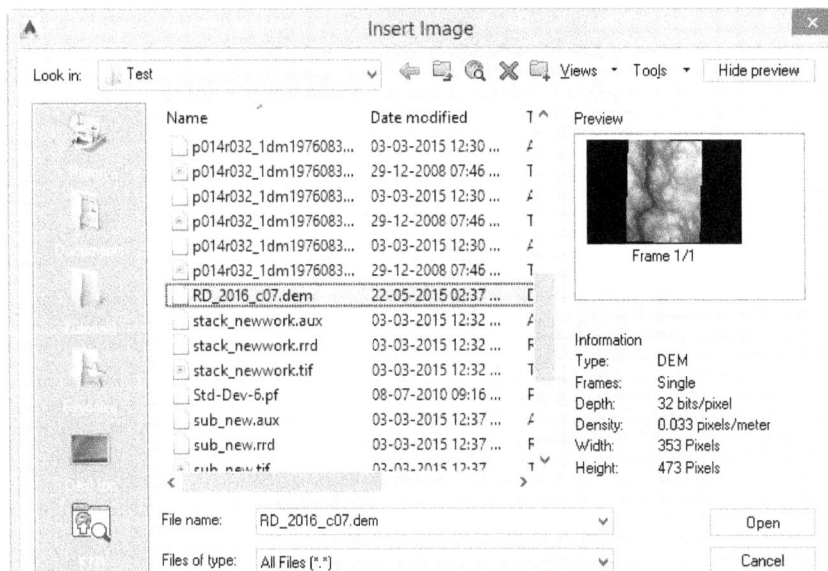

*Figure 7-14 Partial view of the **Insert Image** dialog box*

Select the desired image layers from the **Insert Image** dialog box, refer to Figure 7-14. Ensure that the **Insertion wizard** radio button is selected and then choose the **Open** button; the **Assign Color Map** page will be displayed, as shown in Figure 7-15. The options in the **Assign Color Map** page are discussed next.

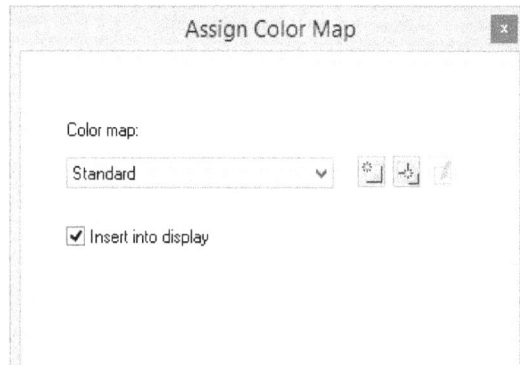

Figure 7-15 *Partial view of the Assign Color Map page*

In this page, you can assign color to the image. You can change the color map any time. In the **Color map** drop-down list, the **Standard** option is selected by default for digital elevation models.

After attaching a DEM file, you can choose the **Create new color map** button or the **Copy color map** button next to the **Color map** drop-down to configure the color map. Note that the **Edit the selected color map** button will be deactivated if a new DEM file inserted in the drawing. If you clear the **Insert into display** check box; the image will not be displayed in the drawing.

If you choose the **Create new color map** button, the **Palette Color Map Definition** dialog box will be displayed, as shown in Figure 7-16.

The **Palette Color Map Definition** dialog box helps to edit or review attributes of a color map in the digital elevation model. Different options in the **Palette Color Map Definition** dialog box are discussed next.

In the **Color map name** edit box, you can enter the desired palette color map name.

The **Data interpretation** drop-down list contains the type of data to be displayed for the surface. By default, the **Value** option is selected in this drop-down list. You can choose any option from this drop-down to display the surface. If you select the **Height (US survey feet)** option from the **Data interpretation** drop-down list, it will display elevation value in feet. The **Height (meters)** option will give elevation value in meters. If you select the **Slope (Percent)** option from the **Data interpretation** drop-down, it will generate slope map of the selected digital elevation model in percentage. The **Slope (Angle)** option generates slope map according to the angle of the slope inclination. If you want to generate aspect map of a DEM region, select the **Aspect** option from the **Data interpretation** drop-down list. This will create an aspect map with direction of the ground slope.

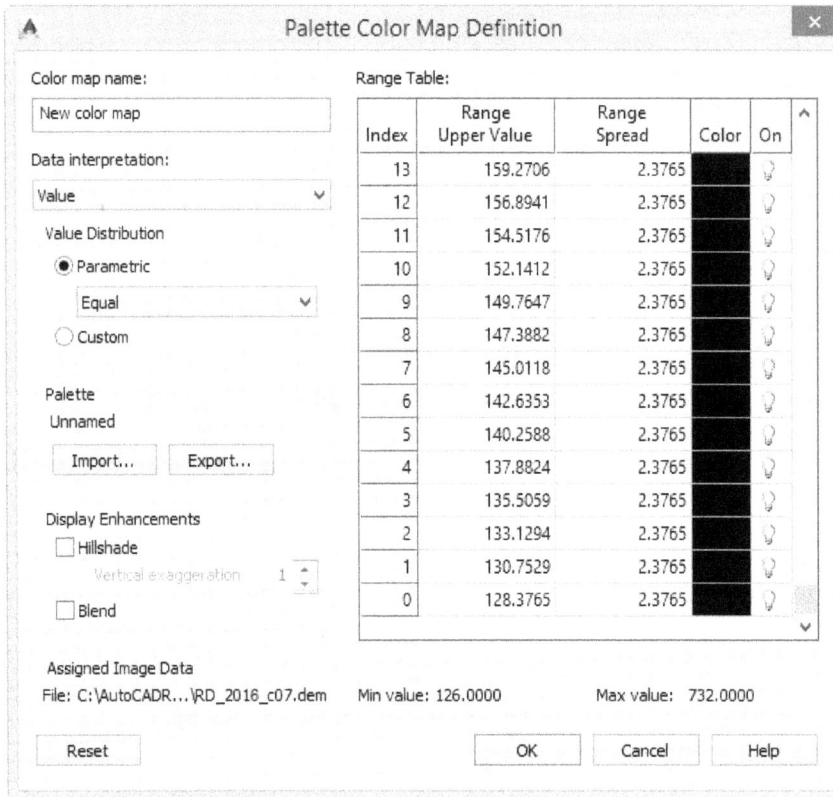

*Figure 7-16 The **Palette Color Map Definition** dialog box*

In the **Value Distribution** area, you can choose the **Parametric** or the **Custom** radio button to set the color distribution in the **Range Table**. By default, the **Parametric** radio button is selected in the **Value Distribution** area. In the **Parametric** drop-down list, you can select the **Equal**, **Standard Deviation**, or the **Quantile** option to distribute the color in the **Range Table**. If you select the **Equal** option, it will distribute each color in an equal range. The upper limit of the third range is the arithmetic mean value of the color in the **Range Table**. The **Quantile** option distributes an equal number of data points in the source file.

If you select the **Custom** radio button, you will be able to set the range spread manually.

In the **Palette** area, the name of the imported color map will be displayed. If you choose the **Import** button, the **Import Palette** dialog box will be displayed, as shown in Figure 7-17.

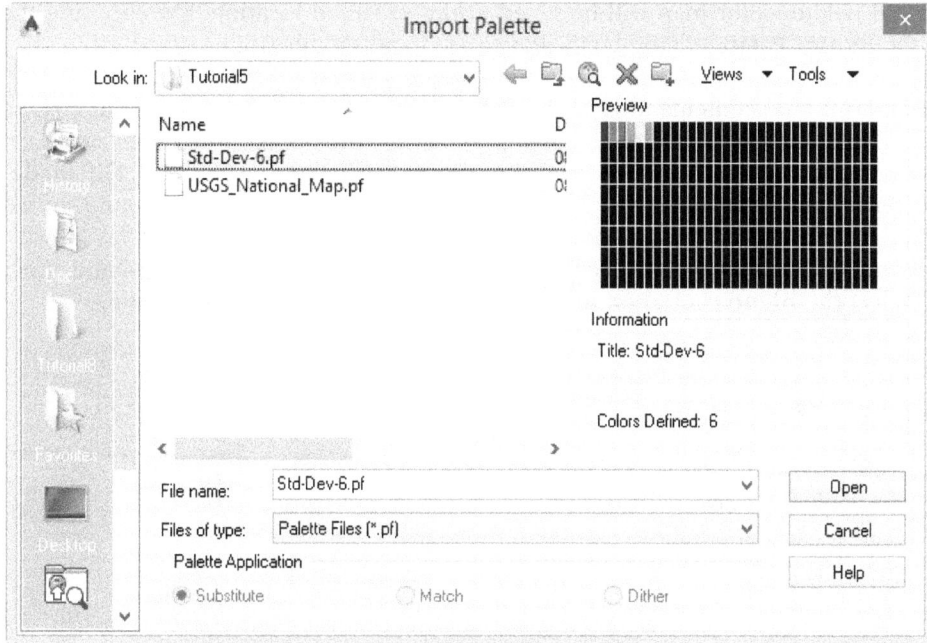

*Figure 7-17 The **Import Palette** dialog box*

You can use the **Import Palette** dialog box to import saved palette to the current image. Some of these options in the window are discussed next.

In the **Look in** drop-down, you can specify the folder or directory that contains the color palette. On selecting the required folder or directory, the contents of the selected folder are displayed in the list box below this drop-down list.

The **File name** drop-down list will display the name of the color palette selected in this window. This drop-down list also displays the path of recently used color palette.

The **Preview** area displays the preview of the selected color palette.

The **Information** area displays the general information such as tile of the selected palette and the colors defined.

After selecting the desired color palette map, choose the **Open** button. The selected color palette map will be imported under the **Color** column of the **Palette Color Map Definition** dialog box.

You can also save the created color palette map. To do so, choose the **Export** button from the **Palette Color Map Definition** dialog box; the **Export Palette** dialog box will be displayed. Enter the desired file name in the **File name** area and choose the **Save** button, as shown in Figure 7-18.

The created palette color map will be saved at the specified location. On choosing the **Save** button, the **Export Palette** dialog box will be closed and the file will be saved.

In the **Display Enhancements** area of the **Palette Color Map Definition** dialog box, you can select the **Hillshade** or the **Blend** check box to change the appearance of the surface. If you want to cast light across the surface from the northwest direction, select the **Hillshade** check box; the **Vertical exaggeration** rollout will be activated. It creates shadows around the valley or hills. You can set the vertical exaggeration level from 1 to 100. If the value is less, it will create light hillshade effects and if you specify more value, it will create intense hillshade effects. You can also select the **Blend** check box to create a smooth transition between colors.

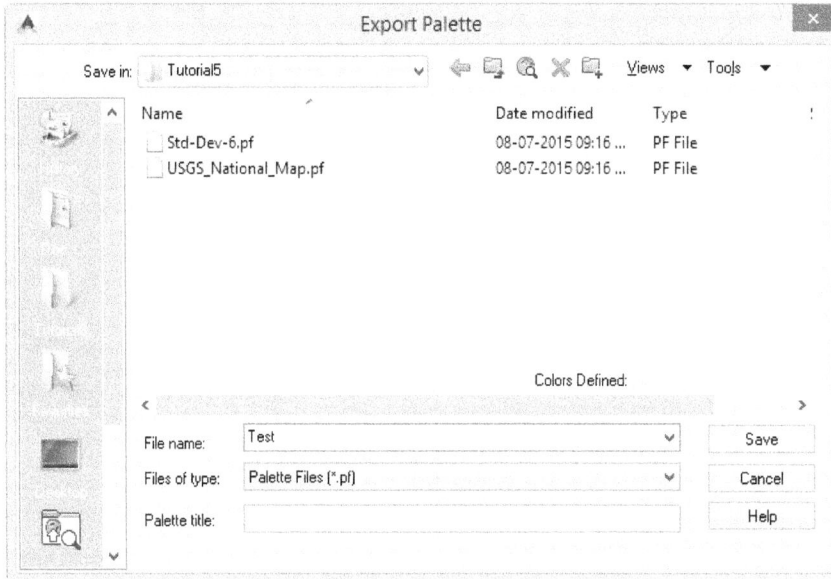

*Figure 7-18 The **Export Palette** dialog box*

The **Range Table** shows the details of color range numbers, range values, colors, and visibility of the color palette map. The **Index** column provides a reference number for each color. You can add or delete rows in the table to set the number of ranges for the color map. The **Range Upper Value** column shows the upper limit of the data values for each range. You can edit these numbers, but they must remain in the correct sequence. The **Range Spread** column shows the size of the range. In the **Color** column, you can also change the color of any of the range column. To do so, double-click on the desired color box, the **Select Color** dialog box will be displayed, as shown in Figure 7-19. Select the desired color from the **Select Color** dialog box to assign the new color in the selected color box. Then, choose the **OK** button; the **Select Color** dialog box will be closed and the selected color will be shown in the desired color box.

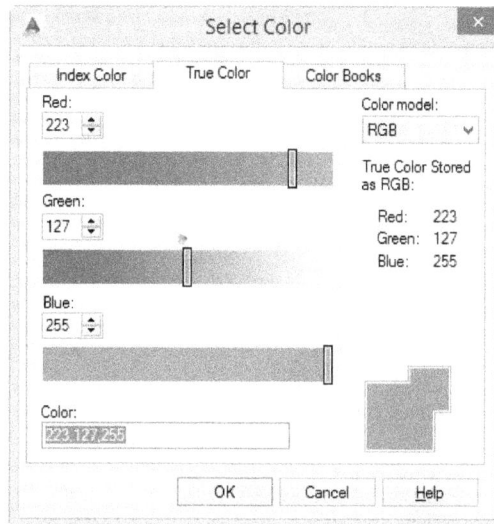

Figure 7-19 *The* *Select Color* *dialog box*

Note

1. You can add or delete the rows from the **Palette Color Map Definition** *dialog box to set the color palette map.*

2. Using the **Palette Color Map Definition** *dialog box, you can create slope and aspect map of a desired DEM region.*

After specifying all the parameters in the **Palette Color Map Definition** dialog box, choose the **OK** button. The **Palette Color Map Definition** dialog box will be closed and the created color palette name will be displayed in the **Color map** drop-down list of the **Assign Color Map** page. Choose the **Next** button; the **Pick Correlation Source** page will be displayed which has been discussed earlier. Choose the **Next** button in this page; the **Modify Correlation Values** page will be displayed, as discussed earlier. Choose the **Next** button in the **Modify Correlation Values** page; the **Insertion** page will be displayed. Choose the **Finish** button in the **Insertion** page; the **Insertion** page will be closed and the modified digital elevation model will be inserted in the window, as shown in Figure 7-20.

Figure 7-20 *The modified digital elevation model*

Note
If the image is not displayed in the drawing, use the **Zoom extents** *option from the Navigation Bar to display the image in the drawing area.*

TUTORIALS
General instructions for downloading tutorial files:
Before starting the tutorial, you need to download and save the tutorial files on your computer. To do so, follow the steps given next:

1. Download the *c07_rd_2016_tut.zip* file from *http://www.cadcim.com*. The path of the file is as follows:
    Textbooks > Civil/GIS > AutoCAD Raster Design> Exploring AutoCAD Raster Design 2016.

2. Now, save and extract the downloaded folder at the following location:

    C:\AutoCADRasterDesign2016

Notice that *c07_rd_2016_tut* folder is created within the *AutoCADRasterDesign2016* folder.

Tutorial 1   Displaying FCC of a Multispectral Image

In this tutorial, you will insert a multispectral image with different bands into the drawing using the **Insert** tool. You will then change the band combinations to create false color composition of the inserted multispectral image.   **(Expected time: 30 min)**

The following steps are required to complete this tutorial:

a. Start AutoCAD Raster Design and insert a multispectral image.
b. Assign band combination.
c. Save the drawing.

Starting AutoCAD Raster Design and Opening the New Template

1. Start AutoCAD Raster Design application and choose **New** from the Application Menu; the **Select Template** dialog box is displayed.

2. Select the **acad.dwt** template and choose the **Open** button to open the template file.

Inserting the Multispectral Image

In this section, you will insert a Landsat MSS image into the drawing using the **Insert** tool in AutoCAD Raster Design.

1. Choose the **Insert** tool from the **Insert & Write** panel of the **Raster Tools** tab; the **Insert Image** dialog box is displayed.

2. In this dialog box, select the **Insertion wizard** radio button in the **Insert Options** area.

3. Next, browse to the location *C:\AutoCADRasterDesign2016\c07_rd_2016\c07_tut01* and then select the following files:
    p014r032_1dm19760831_z18_10.tiff
    p014r032_1dm19760831_z18_20.tiff
    p014r032_1dm19760831_z18_30.tiff
    p014r032_1dm19760831_z18_40.tiff

    Also, ensure that the **Treat as multispectral** check box is selected by default in the **Insert Image** dialog box.

4. Next, choose the **Open** button; the **Insert Image** dialog box is closed and the **Assign Color Map** page is displayed, as shown in Figure 7-21. Ensure that the **Insert into display** check box is selected by default in this page.

5. In the **Assign Color Map** page, choose the **Next** button; the **Pick Correlation Source** page is displayed, as shown in Figure 7-22.

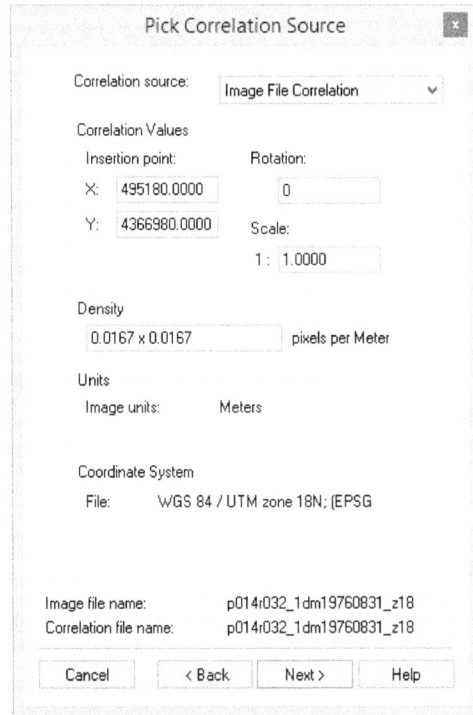

Figure 7-21 The *Assign Color Map* page   *Figure 7-22* The *Pick Correlation Source* page

6.   In the **Pick Correlation Source** page, choose the **Next** button; the **Modify Correlation Values** page is displayed.

7.   In this page, choose the **Next** button; the **Insertion** page is displayed.

8.   In this page, choose the **Finish** button; the **Insertion** page is closed and the image is inserted in the drawing, as shown in Figure 7-23.

Figure 7-23 *Image inserted in the drawing*

Assigning Band Combination

In this section, you will assign false color composition to the inserted image.

1. Choose the **Expand Tree** button from the **RASTER DESIGN** palette; the hierarchy of objects is displayed in the **Tree** view. Ensure that the **Image Insertions** option is selected from the **Image View** drop-down list in the **RASTER DESIGN** palette.

2. Next, right-click on the image; a shortcut menu is displayed, refer to Figure 7-24.

3. Choose the **Edit Color Map** option from the shortcut menu; the **Band Assignment Color Map** dialog box is displayed, as shown in Figure 7-25.

4. In the **Band Assignment Color Map** dialog box, specify the bands in the **Red**, **Green** and **Blue** spinners, as shown in Figure 7-26.

5. Next, choose the **OK** button; the **Band Assignment Color Map** dialog box is closed and false color composition (FCC) is assigned to the image, as shown in Figure 7-27.

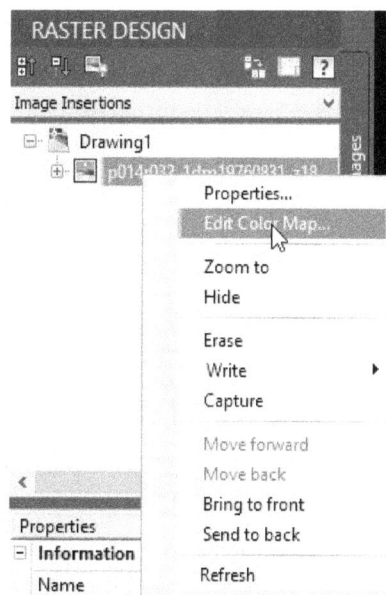

Figure 7-24 *Shortcut menu displayed in the **Image Insertions** view of the **RASTER DESIGN** palette*

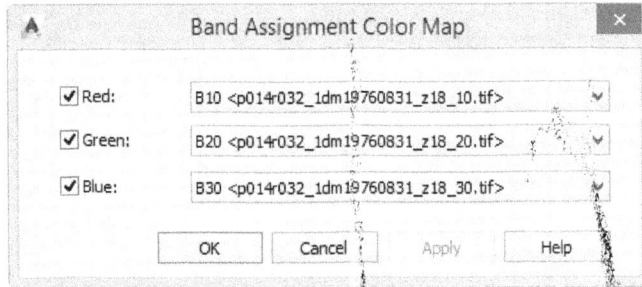

Figure 7-25 The **Band Assignment Color Map** *dialog box*

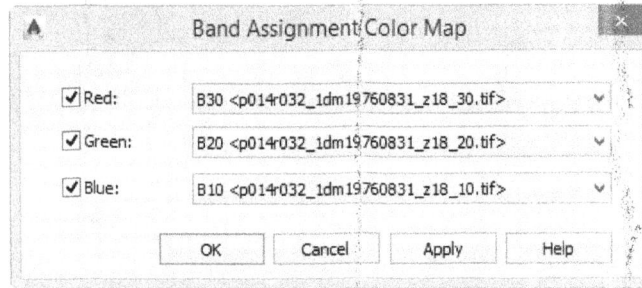

Figure 7-26 *Bands specified in the* **Band Assignment Color Map** *dialog box*

Figure 7-27 *The False Color Composition assigned to the image*

Note

In the FCC, forest cover is visible dark red color, vegetation is in light red, and water body is in blue.

Saving the File

1. Choose the **Save As > AutoCAD Drawing** option from the Application Menu; the **Save Drawing As** dialog box is displayed.

2. In this dialog box, browse to the following location:

    C:\AutoCADRasterDesign2016\07_rd_2016_tut\c07_tut01

3. In the **File name** edit box, enter **c07_Tut01_Result**.

4. Choose the **Save** button; the dialog box is closed and drawing file is saved with the name **c07_Tut01_Result.dwg** at the specified location.

Closing the File

1. Choose the **Close** option from the Application Menu; the file is closed. Ignore the message box displayed, if any.

Tutorial 2   Creating Slope Map

In this tutorial, you will create slope map from a digital elevation model (DEM) and then apply hillshade effects to it.   **(Expected time: 30 min)**

The following steps are required to complete this tutorial:

a. Start AutoCAD Raster Design and insert the DEM file.
b. Create a slope map from DEM and apply hillshade effects.
c. Save the drawing.

Starting AutoCAD Raster Design and Opening the New Template

1. Start AutoCAD Raster Design application and choose **New** from the Application Menu; the **Select Template** dialog box is displayed.

2. Select the **acad.dwt** template and choose the **Open** button to open the template file.

Inserting the DEM File and Creating the Slope Map

In this section, you will insert grayscale image into your drawing for image enhancement.

1. Choose the **Insert** tool from the **Insert & Write** panel of the **Raster Tools** tab; the **Insert Image** dialog box is displayed.

2. In this dialog box, select the **Insertion wizard** radio button from the **Insert Options** area.

3. Next, browse to the location *C:\AutoCADRasterDesign2016\c07_rd_2016_tut\c07_tut02* and then select the **RD_2016_c07.dem** file.

4. Next, choose the **Open** button; the **Insert Image** dialog box is closed and the **Assign Color Map** page is displayed, as shown in Figure 7-28. Ensure that the **Insert into display** check box is selected by default in this page.

Figure 7-28 The Assign Color Map page

5. In the **Assign Color Map** page, choose the **Create new color map** button in this page; the **Palette Color Map Definition** dialog box is displayed.

6. In the **Palette Color Map Definition** dialog box, enter **Slope Map** in the **Color map name** area.

7. Next, select the **Slope (Angle)** option from the **Data interpretation** drop-down list, refer to Figure 7-29.

8. Choose the **Import** button from this dialog box; the **Import Palette** dialog box is displayed.

9. Next, browse to the location *C:\AutoCADRasterDesign2016\c07_rd_2016_tut\c07_tut02* and then select the **Std-Dev-6.pf** file.

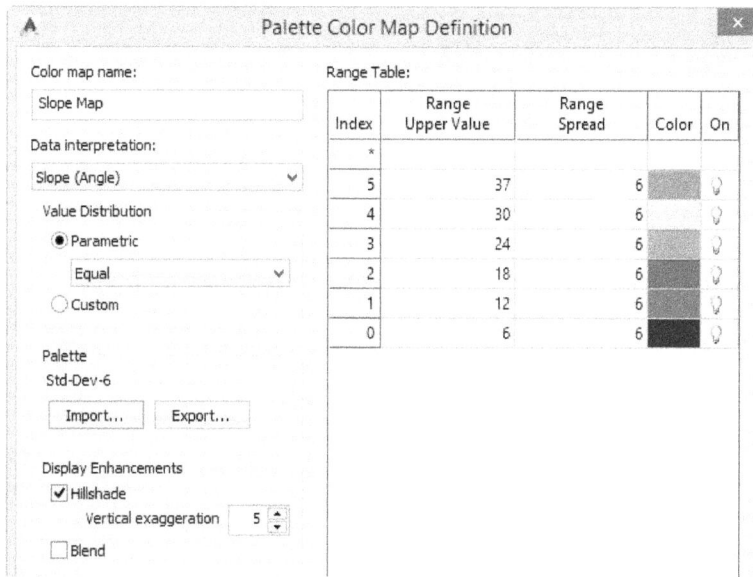

*Figure 7-29 Partial view of the **Palette Color Map Definition** dialog box*

10. Choose the **Open** button; the **Import Palette** dialog box is closed. Ensure that the imported file name is displayed in the **Palette** area of the **Palette Color Map Definition** dialog box.

11. Select the **Hillshade** check box in the **Display Enhancements** area; ensure that the **Vertical exaggeration** edit box gets highlighted.

12. Next, set the value **5** in the **Vertical exaggeration** spinner.

13. Choose the **OK** button; the **Palette Color Map Definition** dialog box is closed.

14. Ensure that the **Slope Map** option is selected in the **Color map** drop-down list of the **Assign Color Map** page.

15. Choose the **Next** button in this page; the **Pick Correlation Source** page is displayed.

16. Choose the **Next** button; the **Modify Correlation Values** page is displayed.

17. Next, choose the **Next** button; the **Insertion** page is displayed.

18. Choose the **Finish** button; the **Insertion** page is closed and the created **Slope Map** is displayed in the drawing.

Note
*If you are not able to see the **Slope Map**, enter **Z** in the Command prompt area and press ENTER. Then, choose the **Extent** option in the Command prompt area; the image will be displayed in full extent.*

Saving the File

1. Choose the **Save As** option from the **Application Menu**; the **Save Drawing As** dialog box is displayed.

2. In this dialog box, browse to the following location:

   C:\AutoCADRasterDesign2016\07_rd_2016_tut\c07_tut02

3. In the **File name** edit box, enter **c07_Tut02_Result**.

4. Choose the **Save** button; the dialog box is closed and drawing file is saved with the name **c07_Tut02_Result.dwg** at the specified location.

Closing the File

1. Choose the **Close** option from the Application Menu; the file is closed. Ignore the message box displayed, if any.

Self-Evaluation Test

Answer the following questions and then compare them to those given at the end of this chapter:

1. Which of the following tools is used to insert a multispectral image into the drawing?

   (a) **Insert**   (b) **Connect**
   (c) **Attach**   (d) None of these

2. Which of the following check boxes is activated after selecting multiple images in the **Insert Image** dialog box?

   (a) **Show frames only**   (b) **Zoom to image(s)**
   (c) **Treat as multispectral**   (d) **All of the above**

3. Which of the following dialog boxes is displayed if the drawing is already set into different coordinate systems?

   (a) **Pick Correlation Source**   (b) **Transform**
   (c) **Modify Correlation Values** (d) None of these

4. The radiometric resolution of IRS LISS III is:

   (a) **6-bits**   (b) **7-bits**
   (c) **8-bits**   (d) None of these

5. The temporal resolution of IRS LISS III is:

   (a) **21 days**   (b) **24 days**
   (c) **25 days**   (d) None of these

6. To edit band combinations, you need to choose the _____ option from the shortcut menu.

7. To edit existing color map of a DEM, choose the _____ button in the **Assign Color Map** page.

8. You can pass infrared band in the _____ channel to create FCC image.

Review Questions

Answer the following questions:

1. On selecting the **Hillshade** check box in the **Palette Color Map Definition** dialog box, the _____ option gets activated.

2. You can import existing color palette map by choosing the _____ button in the **Palette** area of the **Palette Color Map Definition** dialog box.

3. Hillshade effect gives _____ information of the DEM surface.

4. On choosing the **Export** button in the **Palette Color Map Definition** dialog box the _____ dialog box will be displayed.

5. Slope map shows elevation values in _____ .

6. The **Aspect** option is activated on selecting the _____ drop-down list in the **Palette Color Map Definition** dialog box.

7. To create new color map from the DEM data select the _____ button in the **Assign Color Map** page.

8. You can create slope map in percentage in AutoCAD Raster Design. (T/F)

9. You can create multispectral image in AutoCAD Raster Design. (T/F)

10. AutoCAD Raster Design can read *.*tiff* file DEMs. (T/F)

EXERCISES
Exercise 1

Download the *c07_rd_2016_exr.zip* from *http://www.cadcim.com* and create a false color composition of the given image.   **(Expected time: 30 min)**

Exercise 2

Download the *c07_rd_2016_exr.zip* from http://www.cadcim.com and create a slope map of the given digital elevation model, refer to Figure 7-30.   **(Expected time: 30 min)**

Figure 7-30 Digital elevation model

Answers to Self-Evaluation Test
1. a, **2.** c, **3.** b, **4.** b, **5.** c, **6. Edit Color Map**, **7. Create new color map**, **8. Red**

Index

Other Publications by CADCIM Technologies

The following is the list of some of the publications by CADCIM Technologies. Please visit *www.cadcim.com* for the complete listing.

Oracle Primavera Textbook
- Exploring Oracle Primavera P6 v7.0

AutoCAD Map 3D Textbooks
- Exploring AutoCAD Map 3D 2016, 6th Edition
- Exploring AutoCAD Map 3D 2015
- Exploring AutoCAD Map 3D 2014

AutoCAD Civil 3D Textbooks
- Exploring AutoCAD Civil 3D 2016, 6th Edition
- Exploring AutoCAD Civil 3D 2015, 5th Edition

Autodesk Revit Architecture Textbooks
- Autodesk Revit Architecture 2016 for Architects and Designers, 12th Edition
- Autodesk Revit Architecture 2015 for Architects and Designers, 11th Edition

Autodesk Revit Structure Textbooks
- Exploring Autodesk Revit Structure 2016, 6th Edition
- Exploring Autodesk Revit Structure 2015, 5th Edition

Autodesk Navisworks Textbooks
- Exploring Autodesk Navisworks 2016, 3rd Edition
- Exploring Autodesk Navisworks 2015

Autodesk Revit MEP Textbooks
- Exploring Autodesk Revit MEP 2016, 3rd Edition
- Exploring Autodesk Revit MEP 2015
- Exploring Autodesk Revit MEP 2014

ANSYS Textbooks
- ANSYS Workbench 14.0: A Tutorial Approach
- ANSYS 11.0 for Designers

CATIA Textbooks
- CATIA V5-6R2015 for Designers, 13th Edition
- CATIA V5-6R2014 for Designers, 12th Edition
- CATIA V5-6R2013 for Designers, 11th Edition

AutoCAD LT Textbooks
- AutoCAD LT 2016 for Designers, 11th Edition
- AutoCAD LT 2015 for Designers, 10th Edition

AutoCAD Electrical Textbooks
- AutoCAD Electrical 2016 for Electrical Control Designers, 6th Edition
- AutoCAD Electrical 2015 for Electrical Control Designers
- AutoCAD Electrical 2014 for Electrical Control Designers
- AutoCAD Electrical 2013 for Electrical Control Designers
- AutoCAD Electrical 2012 for Electrical Control Designers

SolidWorks Textbooks
- SOLIDWORKS 2016 for Designers, 14th Edition
- SOLIDWORKS 2016: A Tutorial Approach, 3rd Edition
- SolidWorks 2015 for Designers
- Learning SolidWorks 2012: A Project based Approach

CADCIM Technologies Textbooks Translated in Other Languages

Pro/ENGINEER Textbooks
- Pro/ENGINEER Wildfire 4.0 for Designers (Korean Edition)
  HongReung Science Publishing Company, South Korea
- Pro/ENGINEER Wildfire 3.0 for Designers (Korean Edition)
  HongReung Science Publishing Company, South Korea

AutoCAD Textbooks
- AutoCAD 2006 (Russian Edition)
  Piter Publishing Press, Russia
- AutoCAD 2005 (Russian Edition)
  Piter Publishing Press, Russia
- AutoCAD 2000 Fondamenti (Italian Edition)
- AutoCAD 2000 Tecniche Avanzate (Italian Edition)
- AutoCAD 2000 (Chinese Edition)

3D Studio MAX and VIZ Textbooks
- Learning 3ds max5: A Tutorial Approach
  (Complete manuscript available for free download on www.cadcim.com)
- Learning 3ds Max: A Tutorial Approach, Release 4
  Goodheart-Wilcox Publishers (USA)

AutoCAD Textbooks Authored by Prof. Sham Tickoo and Published by Autodesk Press
• AutoCAD: A Problem-Solving Approach: 2013 and Beyond
• AutoCAD 2012: A Problem-Solving Approach
• AutoCAD 2011: A Problem-Solving Approach
• AutoCAD 2010: A Problem-Solving Approach
• Customizing AutoCAD 2010

3ds Max Design Textbooks
• Autodesk 3ds Max Design 2015: A Tutorial Approach
• Autodesk 3ds Max Design 2014: A Tutorial Approach

3ds Max Textbooks
• Autodesk 3ds Max 2016: A Comprehensive Guide, 16th Edition
• Autodesk 3ds Max 2015: A Comprehensive Guide, 15th Edition

Autodesk Inventor Textbooks
• Autodesk Inventor 2016 for Designers,16th Edition
• Autodesk Inventor 2015 for Designers,15th Edition
• Autodesk Inventor 2014 for Designers
• Autodesk Inventor 2013 for Designers

CINEMA 4D Textbooks
• MAXON CINEMA 4D R17 Studio: A Tutorial Approach, 3rd Edition
• MAXON CINEMA 4D R16 Studio: A Tutorial Approach

Flash Textbooks
• Adobe Flash Professional CC 2015: A Tutorial Approach
• Adobe Flash Professional CS 6: A Tutorial Approach

Maya Textbooks
• Autodesk Maya 2016: A Comprehensive Guide, 8th Edition
• Autodesk Maya 2015: A Comprehensive Guide, 7th Edition

Softimage Textbooks
• Autodesk Softimage 2014: A Tutorial Approach
• Autodesk Softimage 2013: A Tutorial Approach

NukeX Textbook
• The Foundry NukeX 7 for Compositors

Fusion Textbooks
• Blackmagic Design Fusion 7.0 Studio: A Tutorial Approach
• The eyeon Fusion 6.3: A Tutorial Approach

Computer Programming Textbooks
• Learning Oracle 11g: A PL/SQL Approach
• Learning ASP.NET AJAX

Paper Craft Book
• Constructing 3-Dimensional Models: A Paper-Craft Workbook

Coming Soon from CADCIM Technologies
• Exploring RISA 3D
• NX Nastran 9.0 for Designers
• Exploring ETabs 2016
• Exploring Primavera P6 v8.4
• Exploring Autodesk Revit Architecture 2017 for Architects and Designers, 13th Edition
• Exploring Autodesk Revit Structure 2017, 7th Edition
• Exploring Autodesk Revit MEP 2017, 4th Edition

Online Training Program Offered by CADCIM Technologies
CADCIM Technologies provides effective and affordable virtual online training on various software packages such as CAD/CAM/CAE, Animation, Civil, GIS, and computer programming languages. The training will be delivered 'live' via Internet at any time, any place, and at any pace to individuals, students of colleges, universities, and training centers. For more information, please visit the following link: *http://www.cadcim.com*

www.ingramcontent.com/pod-product-compliance
Lightning Source LLC
Chambersburg PA
CBHW061407210326
41598CB00035B/6133